AIRCRAFT STRUCTURES AND SYSTEMS

D0294537

AIRCRAFT STRUCTURES AND SYSTEMS

SECOND EDITION

RAY WILKINSON
University of Hertfordshire

MechAero

Published by
MechAero Publishing
PO Box 196
St Albans
AL1 4FD
UK

First edition published by Addison Wesley Longman Limited, 1996. ISBN 0-582-27939-9

This edition published by MechAero Publishing, 2001

ISBN 0-954-07341-X

Set by Pantek Arts Ltd, Cornwallis House, Pudding Lane,
Maidstone, ME14 1NY, England
Printed in Great Britain by Selwood Printing Ltd, Edward Way,
Burgess Hill, West Sussex, RH15 9UA, England

CONTENTS

CHAPTER 12 – AUTOPILOT, RADAR AND RELATED SYSTEMS   136

CHAPTER 13 – AIRCRAFT SYSTEMS   147

CHAPTER 14 – WEAPON SYSTEMS   164

CHAPTER 15 – THE COCKPIT   183

PREFACE

There are many excellent books around that explain how the structure of an aircraft is designed to support the loads that are created in flight and on the ground. My idea when writing this book was to put together something which presents a basic treatment of this complex subject without the formulae, in a form that is easy to browse, and to apply a similar treatment to the systems used in aircraft. I hope that it will appeal to students, apprentices and experienced engineers alike in the aerospace industry, and to Air Cadets, since it covers much of the Airframes syllabus. It will also be useful to pilots and others with a related interest in the way an aircraft works.

Mainly, I hope that this digestible form will satisfy the need and curiosity of many different people in different ways, and to generate an interest and excitement about aircraft in general and aeronautical engineering in particular. Readers are encouraged to look at aircraft in a different light, to appreciate not only their natural grace and elegance (at least most of them), but also the elegance and quality of the engineering, where economy of materials and lightness reign supreme.

The aerospace industry is an extremely volatile one, subject to the boom and bust of the world economy and changing defence needs and strategies probably more than any other industry. With all that, however, it is a fascinating industry to be associated with, and offers many rewards to those prepared to accept its fickle nature.

ACKNOWLEDGEMENTS

I would like to thank Dick Barnard for his advice and encouragement as I wrote this book, and my wife Tina for her patience and understanding.

I would also like to thank Headquarters Air Cadets for their help, and for allowing me to follow the format of the Air Cadet Publication on Airframes.

Many other colleagues and associates helped me along the way, in particular Tony Hindmarsh (Monarch Aircraft Engineering), Martyn Pressnell, Peter Webster, Terry Newman, Sue Attwood (all University of Hertfordshire), Mr M Groeneveld (Sabena Airlines), Mike Tagg (RAF Museum, Hendon), Jim Flaherty (British Airways), Mr Roger Smart (Cobham plc).

Many of the photographs in this book were provided by Alistair Copeland, with additional photographs and illustrations kindly supplied by several companies. The author and publishers are grateful to the following for permission to reproduce copyright material:

Marshall Aerospace for our Figs 3.8 & 8.5; Chris Miller for our Fig. 4.1; British Aerospace Airbus for our Figs 4.8 & 6.9; Rolls-Royce plc for our Fig. 9.2; Raytheon Corporate Jets, Inc. for our Figs 9.8, 11.2 & 13.5; Messier-Dowty for our Figs 10.10 & 10.11; Dunlop Ltd, Aircraft Tyres Division for our Fig. 10.16; The Boeing Company's Maintenance Training organization for our Figs 11.1 & 11.7; British Aerospace (Systems & Equipment) Limited for our Fig. 12.12; Collins Commercial Avionics for our Fig. 12.4; Trimble Navigation for our Fig. 12.6; British Aerospace Dynamics for our Figs 1.1, 14.1 & 14.3; British Aerospace Defense Ltd, Royal Ordnance Division for our Fig. 14.2; Irvin Aerospace Ltd for our Fig. 14.5; Hunting Engineering Ltd for our Figs 14.7 & 14.8; Martin-Baker Aircraft Co. Ltd for our Figs 15.13 & 15.18; Smiths Industries Aerospace for our Figs 15.4, 15.5 & 15.7–15.10; GEC-Marconi Avionics for our Figs 15.11 & 15.12; Helmet Integrated Systems Ltd for our Fig. 15.15.

GLOSSARY

aft — near or in the direction of the rear of the aircraft

angle of attack (α) — the angle between the chord line of a wing or other aerodynamic surface and the oncoming air

ASI — air-speed indicator

aspect ratio — the ratio of wing span to average chord, an indication of the slenderness of a wing

beam — a structural member loaded at an angle (often at right angles) to its length

bending moment — the product of a force and its moment arm

bonding — method of attachment using adhesives

canard — an arrangement of foreplanes and wing, rather than the conventional wing and tailplane (after the French for 'duck')

cantilever — a beam supported only at one end

chord — the distance between the leading and trailing extremities of a wing section

chord line — an imaginary line joining the leading and trailing extremities of a wing section

composite — containing more than one component (in particular materials containing a mixture of plastics and metal, glass or other fibres)

DME — distance-measuring equipment

drag — a force acting aft as a result of the motion of a body through the air

fail safety	principle of maintaining adequate performance after some degree of damage or degradation has occurred (see redundancy)
foreplane	a horizontal stabilising and control surface forward of the wing
flutter	an oscillation caused by the interaction between structural and aerodynamic effects
frame	a hoop-shaped fuselage member which gives it its cross-sectional shape
ILS	instrument-landing system
inertia	a body's resistance to a change in its motion as a result of an applied acceleration
Kevlar	proprietary name for an aramid fibre used in composite materials
lift	a force at right angles to a body's motion through the air, generated as a result of a pressure difference between opposite surfaces
moment arm	the perpendicular distance from the line of action of a force to the point at which the moment acts
redundancy	the provision of alternative load paths or functional routes such that the failure of one element will not cause collapse of the entire structure or total system failure
rib	part of the wing structure which provides the wing-section's shape and supports the skin and stringers
root	the end of the wing closest to the fuselage
shear (force)	a form of loading which tends to cause the atoms or molecules of a material to slide over each other, similar to the action created by a pair of scissors
shock wave	an area of rapid change of air pressure created when air flows at a speed greater than the local speed of sound
spar	a spanwise beam in a wing which carries the majority of bending moment generated by lift, weight and inertia loads
strength-to-weight ratio (SWR)	the ratio of a material's static strength to its density

stress	the intensity of loading, given by the applied load divided by the area over which the load acts
stringer	a member which supports a section of aircraft skin to prevent buckling under compression or shear loads
strut	a structural member which is loaded in compression
TAS	true air-speed
thrust	the force causing an aircraft to travel forwards, overcoming the drag force
tie	a structural member which is loaded in tension
tip	the outermost extremity of a wing
VOR	VHF omnidirectional radio range
VSI	vertical speed indicator
web	a structural member loaded in shear in the plane of the member
wing loading	an aircraft's weight (or effective weight if it is manoeuvring) divided by its gross wing area
wing span	the distance from wing-tip to wing-tip

CHAPTER 1

AIRFRAME DESIGN FEATURES

Objectives: to introduce the main airframe components, describe their main design features, and consider the forces acting upon them and the effects of their weight and shape.

Figure 1.1 Panavia Tornado firing Sky Flash medium-range air-to-air missile. Photograph courtesy British Aerospace Dynamics Ltd.

INTRODUCTION

Most people who fly, perhaps when going on holiday, look at an aircraft and see just that – a complete aircraft – without really considering the huge complexity of parts that lie under the skin. Perhaps they don't want to think about it, preferring the bliss of ignorance at a stressful time. However, others realise that an aircraft is an amazingly complex machine. Just how complex it is can only really be grasped when an understanding of some of the principles has been achieved. Of course, the aircraft is not made so complex just for the sake of it – every component is there for a specific purpose.

The high cost of an aircraft can only be justified if it can provide a return on the money invested. For commercial aircraft this seems quite straightforward – buy the cheapest aircraft that can carry the required number of passengers or amount of freight. But the initial cost of buying the aircraft is usually only a small part of the total cost of ownership throughout its entire operating life. It is often worth paying extra for an aircraft that will be cheaper to operate. When depreciation, operating life and reliability or down time (the time during which the aircraft is unavailable because it is awaiting repair) are taken into account, this can be a difficult decision to make. For military aircraft, the questions are different, and decisions can be more or less straightforward than for a commercial one, especially when political aspects need to be considered.

Whatever the purpose of the aircraft, it must be strong enough and stiff enough, and able to sustain a long life in service, often thirty years or more. It must also be constructed so that if any part fails, as some are bound to do, the failure does not cause the loss of the aircraft, and possibly many lives. There are numerous ways of achieving this, but these are the subject of later chapters. It is important to say, however, that aircraft are among the safest means of transport. This is due in no small measure to the rigorous and detailed requirements applied by aircraft designers and operators, safeguarded by airworthiness authorities world-wide.

THE STRUCTURE

Airframe components

Almost any airframe may be split into four main components:

- the mainplane or wing
- the fuselage or body
- the tail unit (or foreplanes, for a canard-type aircraft)
- mountings for all other systems (undercarriage, engines, etc.)

Each main component is designed to perform a specific task, so that the complete airframe can carry out the job for which it was designed in a safe and efficient way. The main features of each major component are defined in the next few paragraphs, together with a brief description of the loads they will see

Figure 1.2 Airframe major components. *The BAe 146 is a good example of a modern regional jet-transport aircraft. Unusual for an aircraft this size in having four engines, the aircraft's high wing arrangement, popular in this class of aircraft, can be clearly seen here.* Photograph: Alistair Copeland.

during operation. All aircraft are made up of a great many individual parts, and each part has its own specific job to do. But even if it were possible to build an aircraft in one single piece, this would not be the best option. Some parts will become damaged, wear out or crack during service, and provision must be made for their repair or replacement. If a part begins to crack, it is imperative that the structure does not fail completely before it is found during maintenance inspections, or the safe operation of the aircraft may be jeopardised.

The wing

The wing must generate lift from the airflow over it to support the aircraft in flight. The amount of lift required depends on how the aircraft is flying or manoeuvring. For straight and level flight, the total lift produced must be equal to the weight of the aircraft. To take off and climb, the required lift must be developed at a low airspeed. If the aircraft is to fly in very tight turns, the wing must produce lift equal to perhaps eight times the aircraft weight. For landing, the slowest possible forward speed is required, and enough lift must be produced to support the aircraft at these low speeds. For take-off and landing, lift-augmenting devices are normally added to make this possible – flaps, leading-edge slats, etc. The wing needs to be stiff and strong to resist high lift forces, and the drag forces associated with them.

So it could be argued that the wing is the most essential component of an airframe. In fact, aircraft have been designed which consist only of a wing. More commonly, an arrangement that moves some way towards this ideal can be seen in aircraft like the Boeing B-2, F-117 and delta aircraft like Concorde.

In most large aircraft, the wing carries all or most of the fuel, and also supports the main undercarriage; in military aircraft it often carries a substantial part of weapon loads and other external stores. All of these will impart loads onto the wing structure.

The fuselage

The fuselage serves a number of functions:

- It forms the body of the aircraft, housing the crew, passengers or cargo (the payload), and many of the aircraft systems – hydraulic, pneumatic and electrical circuits, electronics.
- It forms the main structural link between the wing and tail or foreplanes, holding them at the correct positions and angles to the airflow to allow the aircraft to fly as it was designed to do. The forces transmitted from these components, particularly the wing and tail, generate a variety of types of load on the fuselage. It must be capable of resisting these loads throughout the required life of the aircraft.
- Engines may be installed inside or attached to the fuselage, and the forces generated can be very high.
- Because of the altitude at which they fly, most modern aircraft have some form of environmental control system (temperature and pressurisation) in the fuselage. The inside of the fuselage is pressurised to emulate a lower altitude than outside, of around 2400 metres (8000 feet) for transport aircraft, and up to 7600 metres (25 000 feet) for military aircraft (with crew

Figure 1.3 F-16. *Blending the fuselage and wing using fairings and fillets reduces drag. In this example, this has been pursued to such an extent that it is difficult to identify the boundary between them.* Photograph: Alistair Copeland.

oxygen), and temperatures are maintained within comfortable limits. These pressure loads generate tensile forces along and around the fuselage, as with the material in an inflated balloon.

These many loading actions can all exist at once, and may vary cyclically throughout the life of the airframe. The fuselage needs to be strong and stiff enough to maintain its integrity for the whole of its design life.

The fuselage is often blended into the wing to reduce drag. In some aircraft it is difficult to see where the fuselage ends and the wing begins.

The tail unit

The tail unit usually consists of a vertical fin with a movable rudder and a horizontal tailplane with movable elevators or an all-moving horizontal tailplane. There is, however, another form of control surface that is finding increasing popularity in fighter aircraft, and even some sport and executive aircraft. In this layout, the horizontal tail surface is replaced or supplemented by moving control surfaces at the *nose* of the aircraft. These surfaces are called *foreplanes*, and this layout is known as the *canard* layout, from the French word for *duck*, which these aircraft resemble.

Whichever layout is used, these surfaces provide stability and control in pitch and yaw, as defined in Figure 1.4. If an aircraft is stable, any deviation from the path selected will be corrected automatically, because aerodynamic effects generate a restoring effect to bring the aircraft back to its original attitude. Stability can be provided artificially, but initially it will be considered

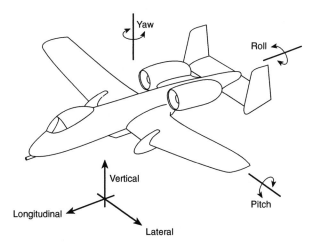

Figure 1.4 Aircraft axes – pitch, roll and yaw. *Motion of an aircraft is defined about three axes, passing through the centre of gravity. Turning about each axis is controlled by a separate set of controls – elevators for pitch control, ailerons for roll and rudder for yaw.*

to be achieved by having a tail unit, with a fixed fin and tailplane, and movable control surfaces attached to them. It is an advantage if the tail is as far from the centre of gravity as possible to provide a large lever – it can then be small and light, with low drag. For this reason it is placed at the rear of the fuselage.

Forces created by the tail act up and down (by the tailplane), and left and right (by the fin). All of these forces, plus the associated bending and torsion loads, must be resisted and absorbed by the fuselage.

WEIGHT

It is good engineering practice for the design of all parts to be as efficient and economical as possible, keeping weight and cost low. Of course, the requirements of low weight and low cost often conflict. In aircraft low weight and high strength are especially important, and great efforts are made at the design stage to achieve this. The maximum weight of an aircraft is set by its design, and any extra weight taken up by the structure is not available for payload or fuel, reducing its operating efficiency. This is made worse by the *weight spiral* effect, where an increase in weight in one area means that other areas need to be strengthened to take the extra loads induced. This increases their weight, and may mean more powerful engines or bigger wings are required to maintain the required performance. In this way, an aircraft may become larger or less efficient purely as a result of poor weight control during design.

There are many ways of saving weight, but one of the most common ones is to use improved materials. Often these may be more expensive, but the extra cost may be justified by the improved performance and reduced operating costs. At the design stage, such questions are the subject of extensive trade-off studies.

AERODYNAMIC FORCES – LIFT AND DRAG

When air flows over aerodynamic surfaces, its pressure will change as its speed changes. This change in pressure, acting over a large area, will produce large forces. Depending on the direction in which these forces act, they will behave as lift or drag forces. Lift is defined as a force at right angles to the direction of flight (i.e. up or down as experienced on the aircraft, with *up* positive) and drag as acting along the direction of flight (i.e. fore and aft, with *aft* positive). If the force is neither exactly at right angles nor exactly along the direction of flight, it will have components of lift *and* drag. As already explained, the lift produced by a large aircraft can be several hundred tonnes, with drag forces typically one tenth of the lift. It is impossible to generate lift without also producing drag, but careful design can ensure that all forms of drag are minimised.

The operator of an aircraft wishes the aircraft to fly at the highest economical speed, to use it to its maximum potential. Commercial operators wish to deliver the maximum cargo or passenger load in the shortest time, and the armed forces gain a great advantage from very fast combat aircraft. The problem they all face is that higher speeds mean higher drag loads. These in turn need more powerful engines, and so increased fuel consumption. The loads increase as the *square* of the airspeed. Inevitably, there is an optimum speed at which to operate any aircraft, and this speed will normally be chosen as the cruise speed for that particular aircraft. To get the best possible performance and best fuel economy an aircraft must be shaped to minimise drag. The external shape has a great influence on the design of the underlying structure. Conversely, it is not possible to design an aircraft to give the minimum drag without taking into account structural factors.

Fortunately, this relationship between load and speed also applies to lift. Lift increases as the square of airspeed. However, the structure must be able to cope with the higher loads generated. Most of the loads that generate the stresses on the airframe structure come about from the effects of aerodynamic pressures on the airframe external surfaces, rather than other sources. These pressures will vary over a wide range, depending on whether the aircraft is cruising, diving, climbing or in turbulent air, and also of course on its speed. Thin wings normally give low drag but may be prone to flutter (a damaging oscillation caused by an interaction between aerodynamic and structural effects); thicker wings are stiffer, and can carry more fuel. Inevitably, the final design is a compromise. The most successful aircraft are those in which the best compromises are found. Figure 1.5 shows examples of high- and low-drag arrangements.

INERTIA FORCES

Inertia forces resulting from manoeuvres may be considerable. If an aircraft is turning, each part of the aircraft will resist any change to its motion by exerting an inertia force on its attachments. The attachments will then exert an equal and opposite force, according to Newton's Laws of Motion. If the aircraft is turning at 4 *g*, i.e. four times the acceleration due to gravity, the component will appear to weigh four times its static or normal weight. This 'weight' must be resisted by the mountings of the component. Thus an engine weighing 15000 N will generate a load of 60000 N in a 4 *g* manoeuvre.

THRUST

The final type of load to consider is the thrust generated by the engine or engines. These loads are transmitted through the engine mounts into the surrounding structure. In constant-speed flight, the thrust will equal the total drag of the aircraft, but usually acts along a different line – the engine centre-line. Pylon-mounted engines are common on large commercial aircraft, and the thrust acting along a line below the wing produces torsional loads on the wing. It is possible to use some loading actions in opposition to other loads on

Figure 1.5 Low- and high-drag arrangements. *Note how streamlined the SR-71 is in comparison with the Fairey Swordfish, showing the importance of low drag for high-speed flight. Streamlining an aircraft usually carries a penalty in terms of higher weight or reduced volume, but is essential in this case.* Photographs: Alistair Copeland.

the structure, thereby reducing the overall strength requirements of the structure and saving weight. This is one of many techniques used by designers to make their aircraft more efficient.

CONCLUSIONS

An airframe must be capable of satisfying many requirements. It must cope with the aerodynamic forces produced at the speeds at which the aircraft is to fly, resist the inertia forces created by the manoeuvres of the aircraft, and of course carry the payload that it was designed to transport. The aircraft must also achieve the other performance aspects for which it was designed, not least to fly at the speeds and with the fuel economy required to provide an effective means of transportation. It must have the lowest possible structural weight, but have a stiff and strong structure with a long life and a high degree of safety. To achieve all of this, designers must resolve many problems. They must have a thorough understanding of the loads on an aircraft structure, and how they are best supported. They must balance the diverse requirements of the operators, the airworthiness authorities, the manufacturing organisation and all of the other specialists, so that the aircraft is safe, economical and effective to operate and to maintain. We can look at the examples in later chapters to see how light yet strong and stiff structures are built.

CHAPTER 2

SHAPE

Objectives: to describe a number of common aircraft shapes, and the advantages, disadvantages and problems arising in choosing a suitable shape; to discuss designing for high flying speeds, with slow landing speeds.

INTRODUCTION

The shape of an aircraft is extremely important, because it dictates how well it can perform a particular function. For a slow-flying aircraft that needs to lift heavy loads, a large wing is needed, together with a light structure. For fast jets, a much smaller wing is required, and the aircraft will be more streamlined. The structure of the aircraft will be much stronger and stiffer, to carry the high loads arising from high speed flight and tight turns. This chapter will describe how aircraft shapes are decided, and other factors that influence the design.

WING LOADING

One of the most important factors in an aircraft design is its *wing loading*, which is simply its total flying weight divided by its wing area. It is usual to take the *gross* wing area, that is the area that includes the part of the wing that is built into the fuselage, because some lift is produced in this area. With simple wings this can be quite straightforward, but when wings are of complex shape it is not always obvious where the boundary of the wing lies, so a degree of judgement is required.

The loads created by the weight of the aircraft vary in flight, depending on the load and fuel it is carrying. Also, manoeuvres require changes in the lift produced, since inertia effects on the aircraft mass need to be overcome (flying at 4 *g* in a turn increases an aircraft's *effective weight* to four times its normal weight, so the lift force produced by the wings must change). But in level flight the aircraft will produce as much total lift (lift from the wing *and* the tail) as its weight, so a useful guide to allow aircraft to be compared is to use the maximum take-off weight (MTOW) to calculate a 'standard' wing loading.

Light aircraft will normally have the lowest wing loading, and fast jets the highest, with transport aircraft in between.

Table 2.1 Wing loading for various aircraft types

Aircraft	Wing area	Max. take-off weight*	Wing loading
Piper Arrow	15.8 m^2	1300 kgf	82.3 kgf/m^2
Hawk 200	16.7 m^2	9100 kgf	545 kgf/m^2
Tornado	25 m^2	28 000 kgf	1120 kgf/m^2
Dornier 328	40 m^2	12 500 kgf	312.5 kgf/m^2
Airbus A300	260 m^2	165 000 kgf	635 kgf/m^2
Boeing 747	511 m^2	350 000 kgf	685 kgf/m^2

* 1 kgf = 9.81 N

MONOPLANES AND BIPLANES

Although there are still a few biplanes around, most aircraft now are monoplanes, having a single wing. This requires a very stiff, strong wing, but avoids the drag penalty of the biplane arrangement. One of the few modern biplanes is the Pitts Special (Figure 2.1); this tiny aircraft uses two wings to give a large wing area with a small wing span, allowing rapid roll rates – its aerobatic performance is remarkable. The two wings are braced together, forming a particularly stiff structure.

Many light aircraft are *braced monoplanes*, having a diagonal bracing strut between the wing and fuselage (Figure 2.2). Without this strut, the wing would need to be stiff enough to resist all of the bending loads created by the lift force on the wings, requiring more structure and hence increased weight. The strut takes some of the lift loads, allowing a lighter structure in the wing, but at the expense of extra drag. Because of the low flying speed of the aircraft, the extra drag caused is small, and therefore acceptable in view of the weight saved.

Because of the drag penalty of bracing struts, the pure cantilever wing is used for all aircraft of medium and high speeds. A cantilever is simply a beam that is supported at only one end – it is explained in more detail in Chapter Four. The cantilever wing arrangement can be categorised as low-wing, mid-wing or high-wing, depending on where it is attached to, or passes through, the fuselage. Typically, the low-wing arrangement seems to be preferred for jet transport and many light aircraft, high wing for turbo-prop transport aircraft and both low- and mid-wing (shoulder-wing) for combat aircraft, but there are many exceptions.

Figure 2.1 Pitts Special. *The Pitts Special is unusual in being one of the few modern biplanes. This very specialised aircraft is extremely agile, capable of outstanding aerobatic performance, due partly to its small wing span in relation to its wing area. The use of the biplane arrangement allows very rapid rolls.* Photograph: Alistair Copeland.

Figure 2.2 Braced monoplane. *The Shorts 360 is a typical example of a braced monoplane. The bracing strut, which links the fuselage to the wing part-way along the span, supports much of the loads imparted by the lift forces on the wing. It allows lighter wing structure at the expense of extra drag.* Photograph: Alistair Copeland.

A cantilever wing must be strong enough and stiff enough to carry the whole weight of the aircraft, and its aerodynamic loads, without the need for external bracing. For a Boeing 747 weighing 350 tonnes, the wing will need to be capable of resisting loads of over 1000 tonnes without failure or excess distortion. This is because manoeuvres and wind gusts cause loads that are several times the aircraft weight. Chapter Six shows how this can be achieved. It must also be able to cope with the highest speeds and manoeuvre loads of the aircraft without deflecting too much, which can cause aerodynamic flutter and may result in collapse or loss of control.

Generally, high speeds require a smaller wing span and low wing area, hence a high wing loading. Conversely, a large span and high wing area, i.e. low wing loading, are best for low speeds. For take-off and landing, it is possible to change the wing area and wing section to some extent by the use of flaps at the trailing edge. This makes the wing structure more complicated, but is desirable or even essential if the aircraft is to land at a safe speed. The design of the wing for a high-speed aircraft, such as the Tornado, will be principally driven by stiffness requirements, to avoid flutter at high speed. High speeds also require minimum drag, so retractable undercarriages and low frontal area are required. Even with a streamlined aircraft, high speeds demand high thrust, and turbo-fan or turbo-jet engines will be needed, in preference to the turbo-prop engines most efficient for lower speeds. At very high speeds, the cross-section of the fuselage and wing together are very carefully designed, using the *area rule* (Figures 2.3 and 2.4) to achieve low drag, and leading to some very complex aircraft shapes. The area rule principle considers the cross-sectional area of the fuselage *plus* wings – if this area corresponds to that of the minimum drag body of similar cross section, then the transonic or supersonic drag will be minimised.

Figure 2.3 Su 27. *The Su 27 is an example of a modern Russian combat aircraft. Its fuselage has a distinctive shape as a result of applying area ruling, which reduces supersonic wave drag and can produce a significant performance increase in supersonic aircraft.* Photograph: Alistair Copeland.

Figure 2.4 Area rule. *As can be seen on this Rockwell B-1, the shape of the fuselage is designed to produce a smooth curve when the total cross-section (fuselage plus wings) is plotted. This creates the characteristic 'Coke-bottle' fuselage shape.* Photograph: Dr R H Barnard

Strength-to-weight ratio

Because high-speed aircraft need small wings for low drag, the loads on these wings will be very high, so the wings will have to be made much stiffer and stronger to carry the wing loads. This leads to increased weight, which the designer tries to avoid. Wing loading is tending to increase over the years, but the designer makes sure that the material is used to best effect, and uses the strongest and lightest materials. In this way, the strength-to-weight ratio of the *structure* is improved. Improved materials can also play a part in allowing higher stresses to be used, and although they may be much more expensive they can save cost by making the design simpler and more efficient. It is important to realise that materials with a high strength-to-weight ratio do not automatically produce a structure with the same qualities. What is important is that the most suitable material is used, together with a simple and effective design. The material must be highly loaded, or it is not being used to best effect, but must not be over-stressed or it will fail early in service.

Stiffness-to-weight ratio

Another important feature of some structures, for instance wings, is the ratio of their stiffness to weight. A wing may be strong enough to withstand the loads upon it, but may lack the stiffness needed to keep its shape accurately in flight. This would be a major problem, and increasing the stiffness may well require the use of more material, increasing weight. In some applications, particularly small components, the materials with the highest strength-to-weight ratio may not be the best to use, because the material may need to be too thin to provide enough stiffness. A good example of this is model aircraft – they use balsa, which is never used structurally in full-size aircraft. If model makers used aluminium alloys, the components would be so thin that they would be very flimsy, and stiffening them up sufficiently would make the models far too heavy to fly. So the stiffness of a structure depends on both its design and the materials used, as will be seen in Chapter Four.

SWEEP BACK, SWING WINGS AND DELTA WINGS

For aircraft flying at or near supersonic speeds, the air flow over the aircraft is very different from that in subsonic flight, and the designer has a new set of problems to face. An aircraft flying through the air generates pressure waves, which move at the speed of sound. At speeds below the local speed of sound these pressure waves 'warn' the oncoming air that the aircraft is approaching. As aircraft speeds approach the speed of sound, these pressure waves no longer travel significantly faster than the aircraft, and a shock wave will begin to form on parts of the aircraft. Air pressure, speed and temperature change very suddenly across this shock wave, and the flow behind it is quite turbulent. The effect on the aircraft is a reduction in lift, an increase in drag, changes in trim and possibly buffeting of the controls. Designers have reduced the effects of these problems with better designs, particularly swept wings. To understand how this is achieved, consider the airflow to be made up of two components – one at right angles to the wing leading edge and one parallel to the leading edge. By sweeping the wings back (or forwards), the component at right angles to the leading edge is reduced as the cosine of the sweep angle. The larger the sweepback angle the higher the speed at which this component will remain subsonic. Highly swept wings give other problems, particularly structural, and are more difficult and expensive to build than simple straight wings. Aerodynamic problems associated with swept wings include an increased likelihood of tip stalling, which can lead to spinning, and a reduced lift-to-drag ratio.

Once flying above the speed of sound, the airflow is changed to become steady again, although quite different to subsonic conditions. Shock waves form whenever there is a change in airflow direction, and curved shapes no longer produce the lowest drag – sharp edges create a single shock wave, and are generally more efficient than curves. The plan form of wings becomes very important, and a low aspect ratio and highly swept wings are most suitable. The reduced aspect ratio of these wings makes life easier for the structural designer, but the loads are very high, and the structure must be strong and stiff.

The main disadvantage of swept wings is that, at a given angle of attack, they produce less lift than an unswept wing of the same general dimensions. This is because the geometry of the sweep angle acts to reduce the *effective* angle of attack. So when the aircraft is flying slowly, for instance during take-off or landing, a larger angle of attack is required to provide sufficient lift. High angles of attack on touchdown and take-off can cause problems with ground clearance at the aft end, and with poor pilot visibility. Concorde is a good example of this, and overcomes the visibility problem in a unique way with its 'droop snoot'. A larger wing could be used – this would also improve the turn performance for a fighter, for example, but would reduce its top speed because of higher drag and weight.

If the sweepback could be changed in flight (swing wing), the aircraft shape could be optimised for a wide range of flying speeds. This has been done on

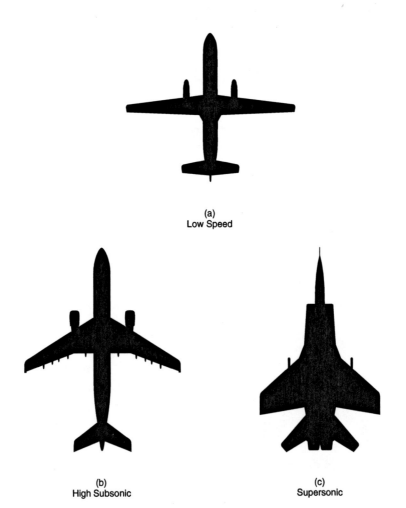

(a)
Low Speed

(b)
High Subsonic

(c)
Supersonic

Figure 2.5 Sweepback for high-speed flight. *Sweepback is used to reduce the undesirable effects caused by high-speed flight. The sweep angle of the wing effectively reduces the speed of the air flow at right angles to the leading edge. The sweep angle required is greater as the maximum speed of the aircraft increases.*

some high-speed military aircraft – in the forward (unswept) position it gives a straight wing of high aspect ratio for good low-speed performance. The forward position allows tight turns at low speeds and also makes flaps more effective for take-off and landing. In the swept position it is highly suited to high-speed flight.

Figure 2.6 Effect of large sweep angle on angle of attack. *Highly swept wings, in this case a delta wing, require high angles of attack to produce sufficient lift for low-speed flight. This can cause difficulties with ground clearance at the rear of the aircraft, and with poor visibility for the crew. Note the very long undercarriage to give the required ground clearance, the upswept rear fuselage and the 'droop-snoot' – a unique solution to the visibility problem.* Photograph: Alistair Copeland.

Figure 2.7 Swing wings. *Swing wings allow the aircraft to offer the best of both worlds – high aspect-ratio unswept wings for good low-speed performance, allowing short take-off runs with large loads; highly swept wings for excellent high-speed performance.* Photographs: Alistair Copeland.

Figure 2.8 Delta wings. *Delta wings are re-emerging on fighter aircraft, as on this Rafale, because they offer the advantages of a large, highly swept wing with high structural stiffness. Many modern fighters use fly-by-wire and advanced computer techniques to great advantage in combination with this wing type.* Photograph: Alistair Copeland.

Another option for aircraft that need to fly at high speeds but also need to be able to turn tightly at all speeds is the delta wing. This has the advantage of high sweepback, but the large chord at the wing root (near the fuselage) gives a low-aspect-ratio wing that is very stiff and strong. They are highly suited to carrying wing-mounted stores, because the wing is also very stiff in torsion. Because the chord of the wing is so large, the wing can be relatively thick, which provides a large volume for storing fuel. Compare this with the English Electric Lightning, with its highly swept wing, which needed to have a large belly tank added because it could carry so little fuel in the wing. It was a very effective interceptor, because of its outstanding climb performance, but even when modified could stay in the air only for a short time.

Because of the aerodynamics of delta wings, they are capable of producing lift at much higher angles of attack than other wing shapes, and so are suited for highly agile fighter aircraft. Delta wings are becoming more common on current combat aircraft, and many examples can be seen, often in conjunction with canard foreplanes for control.

ASPECT RATIO

The aspect ratio of an aircraft's wing is an important design feature, and is simply the ratio of the wing span to its mean aerodynamic chord (the average distance between the leading and trailing edges). Mean aerodynamic chord is not always simple to calculate if a wing shape is complex but, since it is equal to *wing area* divided by *span*, another form of this definition is:

$$\text{aspect ratio} = \frac{\text{span}^2}{\text{wing area}}$$

It is usual to use the *gross area* to calculate the aspect ratio – to include that part of the wing that is inside the fuselage, because this part of the wing produces a certain amount of lift. Typical aspect ratios vary from about 25 for a sailplane (glider) to perhaps 8 for a fighter. High aspect ratio reduces the induced drag caused by air flowing around the wing tips, and is ideal where slow, long-duration flights are required, particularly at high altitude. The drawback is that long, thin wings tend to be heavier, and are quite flexible. An airframe is a 'living structure' so all wings constantly bend and twist in flight, as a result of the aerodynamic loads on them. However, if the aspect ratio is high, and the aircraft flies at a speed that generates high loads, the flexibility of the wings causes unacceptable distortion. In some cases, torsional oscillation of wings can occur, which is one form of *flutter* and leads to high drag and possible structural failure. Every wing will flutter if flown at a high enough speed, so this aspect of flight is very carefully investigated both in the wind tunnel and in early flight trials of new aircraft. In less severe instances, the offset loads from control surfaces such as ailerons can cause the structure to twist significantly. If the structure is not sufficiently stiff, the twisting may be severe enough to cause the movement of the surface to have the opposite effect to that intended, a situation called *control reversal* (Figure 2.9).

THE BEST COMPROMISE

All aircraft designs are a compromise, because it is not possible to get the best possible design in one respect without conflicting with another. The

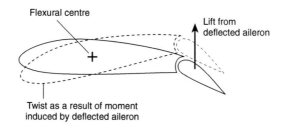

Figure 2.9 Control reversal. *The offset load from an aileron must cause the wing to twist, since it cannot be infinitely stiff. The twist will reduce the effect of the control movement, which is acceptable provided the effect is small. If the structure of the wing is not stiff enough in torsion, the wing will twist sufficiently to increase lift rather than reducing it, and vice versa, reversing the intended effect of the control movements.*

aerodynamicist may want a smooth, thin wing, but the structural designer wants the wing to be thick, to make it stiffer. The weights engineer might prefer there to be no wing at all, since no payload can be carried in it! This is of course an exaggeration, but Figure 2.10 shows how an aircraft might look if it were designed by each of the various specialists. Inevitably, then, the best aircraft is the one in which the designer has found the best set of compromises.

Designing an aircraft is an iterative process – it goes around a number of cycles before the final design is reached. There are many other factors that affect how the aircraft will look – materials availability, available technology,

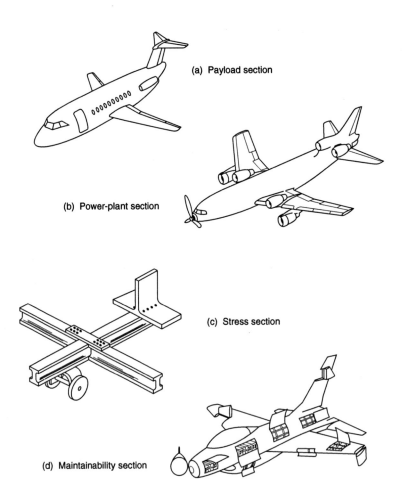

(a) Payload section

(b) Power-plant section

(c) Stress section

(d) Maintainability section

Figure 2.10 Aircraft designed by various specialists. *This set of sketches shows an exaggerated idea of what each of the aircraft design specialists wishes to see in their ideal aircraft. Of course, none of these designs is practical. Rather, they highlight the importance of making sure the design is not driven by any specific requirement other than the real purpose for which it is designed.*

company standard practices and the operating requirements of the user all influence the final design. It is common with many aircraft types for the user to have a strong input to the design specification. After all, the manufacturer is seeking to produce an aircraft that will satisfy the user in all respects.

The designer has many problems in selecting the right shape for an aircraft, and must constantly seek compromises to get the best performance in all the different parts of the aircraft's flight. Once the overall shape of the aircraft has been finalised, the problems of making the airframe, in materials that provide the best strength-to-weight ratio and stiffness, must be tackled. This is the subject of the next chapter.

CHAPTER 3

AIRFRAME LOADS

Objectives: to describe the wide range of loads that can be generated on an airframe, how these loads arise, and give an indication of how the structure may be designed to support these loads.

INTRODUCTION

An aircraft in flight is subjected to an enormous range of different forces and loads. Each of these loads is caused by a different action, but they can be grouped into similar types. Many of the loads act on different parts of the structure, and some can occur in combination with, or as a result of, others. Designers must consider all of these loads, and demonstrate that the aircraft is capable of withstanding them without being damaged, so that it is safe to fly. They must be able to predict what these loads will be, and to prove the strength of the aircraft by structural testing.

TYPES OF LOAD

To understand the way that loads occur on an aircraft, it is necessary to understand the various types of load that can exist. There are quite a number of different ways in which any structure can be loaded. The simplest form of load is a concentrated *force*. This can be created by the weight of an object, i.e. its mass times the acceleration due to gravity, or by an inertia force – its mass times another kind of acceleration, caused by the aircraft manoeuvring. Force can also be generated by some other action – thrust from an engine for example.

A force applied lengthways to a piece of structure will cause *tension* or *compression*, depending on its direction (Figure 3.1). If a force is applied at right angles to a piece of structure, it will generate a *bending moment* (Figure 3.2), which is the force times the moment arm or lever about which it acts. A bending moment, as the name infers, will cause the structure to bend.

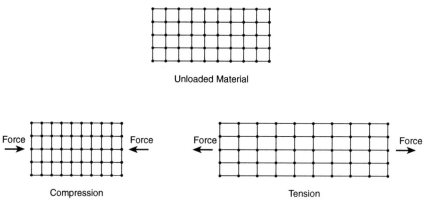

Unloaded Material

Force Force Force Force

Compression Tension

Figure 3.1 Tension and compression. *A force acting along the axis of a piece of structure generates tension or compression, depending on the direction of the force.*

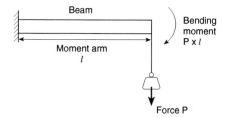

Figure 3.2 Bending moment. *A force acting at an angle to the axis of a beam generates a bending moment, given by the force times the moment arm, measured at right angles to the line of action of the force. The longer the moment arm or the larger the force, the greater the bending moment.*

If a force is offset from the line of a beam as shown in Figure 3.3, it will also cause torsion or twisting of the structure.

The torsion itself does not put the material in either tension or compression, but under a different kind of load – shear. Shear (Figure 3.4) is a form of loading which tries to tear the material, causing the atoms or molecules that make up the material to slide over one another. Note that shear is also created in the case shown in Figure 3.2, because the load is acting at right angles to the beam. Shear is probably the most difficult to visualise, but the action of a pair of scissors (or shears) illustrates the action very well. Each blade forces part of the material in the opposite direction to the other, causing the material to be sheared where the blades meet. Shear is therefore a combination of two actions equal in magnitude and opposite in direction.

This raises an important point – that of *equilibrium*. Equilibrium is the state in which all forces and moments are exactly in balance. Newton's Third Law says that for every action there is an equal and opposite reaction. For every

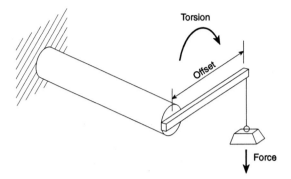

Figure 3.3 Torsion. *A force acting as shown generates torsion, given by the force times the offset measured at right angles to the line of action of the force. The greater the offset, the greater the torsion.*

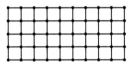

Unloaded Material

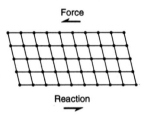

Figure 3.4 Shear. *Two forces, or a force and its reaction, acting as shown generate shear, which is the action utilised by a pair of scissors when cutting. Almost invariably a concentrated force applied to a structure will generate shear.*

force or moment acting on a structure, there is another force or moment holding the structure steady.

As has been seen, each type of load on a structure acts in a different way, although many loads are combined with others. Each load causes a *stress* in the material, which may be defined as the local intensity of loading – the applied load divided by the area of material over which the load acts. Put another way:

$$\text{Stress} = \frac{\text{Load}}{\text{Area over which load acts}}$$

Strain is the proportional deflection within a material as a result of an applied stress. It is important to realise that it is impossible for any structure to be subjected to a stress without also experiencing strain, since the two are inextricably linked. Each form of load will need a different type of structure to carry it most efficiently. With tensile loads, the only critical factor is the area of material that is under stress. In this case it is the cross-sectional area at right angles to the force – within reasonable bounds, the shape of this cross section does not affect the magnitude of the stress. Compressive loads are similar, but can cause the structure to *buckle* (Figure 3.5) – to bend away from the line of action of the applied load. To avoid buckling, the structure needs some material to be placed away from its centre line, which stiffens the structure. A tube is very good at resisting buckling, because the material is concentrated at a uniform distance from the centre line, therefore under equal stress.

Tubes are also very good at carrying torsional loads, which place the wall of the tube in shear. Again, all of the material is at the same distance from the centre line. But shear loads on very thin sections will cause material to buckle if the stress becomes too high. Increasing the thickness reduces the stress, and with it the likelihood of buckling, but this can be achieved without adding much extra weight – Chapter Four will show how *stringers* may be used to support thin skins to prevent buckling under compression or shear.

Pressure loads within fuselages generate *circumferential* or hoop stresses, which load the fuselage skin in tension. On doors and windows, they also create bending loads, as the pressure loads spread across the door are carried by the hinges and latches that hold doors in place, and by window frames.

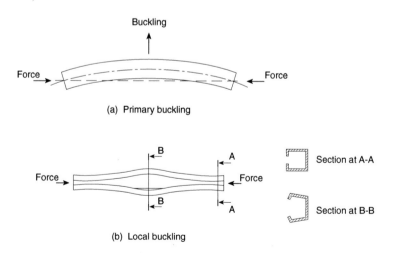

Figure 3.5 Compression and buckling. *Compressive loads may cause the structure to buckle – to bend away from the line of action of the applied load. Open sections, like the 'C'-section shown here, are particularly susceptible to local buckling.*

FUSELAGE LOADS

During the various phases of flight and movement on the ground, the fuselage will experience a wide range of loads from a number of sources. The loads will vary in size, and often in direction, either or both of which may cause problems with *fatigue*, which will be dealt with in detail in Chapter Five. Many loads can occur in combination with others, and the designers must make sure that they have found the worst combination that might occur during the life of the aircraft. They must also satisfy themselves (and the authorities) that the design is capable of supporting these loads with an appropriate margin of safety.

Bending loads

During *straight-and-level* flight, the fuselage is supported by the wing, with another force from the tail that is normally downwards. For canard aircraft, the foreplanes will always produce an upwards force. The weight of the fuselage structure and payload will cause the fuselage to bend downwards from its support at the wing (Figure 3.6), putting the top in tension and the bottom in compression.

In *manoeuvring* flight, the loads on the fuselage will usually be greater than for straight-and-level flight, depending on the manoeuvre flown. During negative *g* manoeuvres (pitching nose-down), and during landing and taxiing,

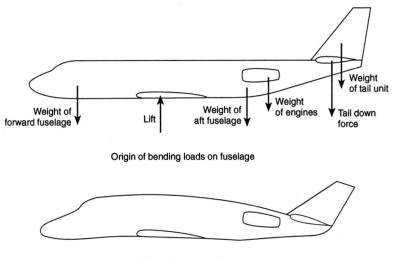

Origin of bending loads on fuselage

Effect (exaggerated)

Figure 3.6 Fuselage bending loads. *Bending loads on the fuselage are created mainly by its own weight and inertia and those of the tail, and tail aerodynamic loads. Fuselage-mounted engines also load the fuselage, as does any payload carried; the further the weight is from the centre of gravity, the higher the bending moments produced.*

some of the loads are reversed, so the structure must be designed to withstand this load reversal. In particular, landing loads may be significant, with a large undercarriage load applied close to the centre of the fuselage and high inertia loads from the nose and tail. The forces caused by the nose-wheel touching down may also be high during a heavy landing.

The bending loads on the fuselage will obviously be higher when the weight is distributed towards the nose and tail, and when the aircraft is heavily laden. In this case, particular attention must be paid to how the payload is distributed in the fuselage. Fuselage-mounted engines also create extra loads on the fuselage. Situating the engines on the wing removes the loads from the fuselage, although of course this is at the expense of higher loads on the wing. Since wing-mounted engines are normally mounted close to the undercarriage, though, where strength is already high, the weight penalty will generally be less.

Pressure loads

Passenger and freighter aircraft are usually pressurised through most of the fuselage. The pressurisation varies with altitude, and is partly controlled by the crew, but sitting inside an aircraft cruising at, say, 9760 m (32 000 feet) passengers would be breathing air at a pressure equivalent to an altitude of about 2440 m (8000 feet). With any physical exertion, they would find it quite difficult to breathe, but sitting relatively still causes no discomfort at all. The cabin altitude is usually changed quite slowly, beginning pressurising long before 2440 m is reached, which is more comfortable, because it causes less ear 'popping' and sinus problems.

Combat aircraft have no need for pressurisation of large areas of the fuselage. In fact, it would generally be a disadvantage, because the aircraft may be subjected to battle damage. Pressurisation is restricted to the cockpit area, plus possibly some electronics bays, where the increased air density may assist with cooling. The cockpit altitude of combat aircraft will be higher than for transport aircraft, perhaps as much as 7620 m (25000 feet), for an aircraft with a ceiling of 15 200 m (50000 feet). This produces a maximum differential of about 26 kPa (3.8 lb per square inch). The low air pressure at this cockpit altitude will not support human life, so the pilot wears a face mask to supply oxygen.

Although pressurisation is used in flight, it also occurs when the aircraft is on the ground. The aircraft may land with the fuselage still slightly pressurised, and ground pressurisation tests will also be carried out from time to time. Underpressure, where the inside is at a lower pressure than the outside air, may also be generated on the ground and in flight, for instance during descent. A relief valve limits underpressure for safety.

The effect of cabin pressurisation is to create loads that try to burst the fuselage. The skin itself carries these loads, putting it in tension. There is also a force stretching the fuselage along its length, which is the pressure difference multiplied by the cross-sectional area of the fuselage. In cylindrical fuselages,

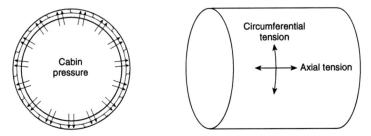

Figure 3.7 Pressurisation loads. *Cabin pressurisation generates stresses in the fuselage skin and structure, which attempt to burst the pressure cabin. These put the skin in tension in all directions in the plane of the skin. For convenience, we consider the tension in two directions – circumferential and axial.*

this load is generated on the front and rear pressure bulkheads, and also on the windscreen and its associated structure.

Particular problems occur below windscreens, and in other areas where the fuselage is required to be non-cylindrical. The internal pressure will generate large bending loads in the fuselage frames, and the structure in these areas must be reinforced to withstand these loads. The aft end of transport-aircraft fuselages is often oval in section, but it is usual to keep the cross-section circular or near-circular until the rear pressure bulkhead is reached.

Depressurisation

Because fuselages are pressurised, the designer must consider what will happen if the pressurisation is lost. This is likely to occur on at least one aircraft of a particular aircraft type at some time. For transport aircraft, it is usually caused by failure of some part of the pressure cabin, frequently a window mounting, by a bird strike on a windscreen, terrorist action or failure of the pressurisation system itself, such as multiple engine failure. In combat aircraft, it is usually by battle damage, bird strike, canopy release or engine failure.

The effects of depressurisation depend on the rate of pressure loss. For relatively slow rates, no structural problems are likely, but the pilot may need to carry out a violent manoeuvre to descend rapidly to a safe altitude. For very high rates (*explosive decompression*), the rapid changes of internal pressure may cause far higher loads than would occur during normal operation. The excessive loads could result in bursting open of overhead bins and other closed compartments, and a large amount of flying debris. In extreme cases, there may be collapse or distortion of floors and internal bulkheads. Because air is compressible, considerable energy is stored when it is compressed, and if released suddenly this energy can cause great damage.

In combat aircraft, cockpit pressurisation is much lower, mainly because depressurisation is more likely. The effects are not usually important structurally, but may affect the crew. Effects on the crew may include oxygen deficiency and hypothermia, although protection is provided by the oxygen mask and aircrew clothing. The ambient pressure of even pure oxygen near maximum operating altitudes is too low to maintain consciousness for more than a few seconds, unless the oxygen is supplied under pressure.

Effects of cut-outs

Doors, windows and hatches are a major problem in the design of the aircraft, because they interrupt the load-bearing structure of the fuselage. Windows, being small, do not create a severe problem, and the structure around them is reinforced, with windows themselves not considered to support any fuselage loads. Doors may or may not carry some of the load in the fuselage structure, depending on their design. With passenger doors, it is necessary to be able to open them quickly and easily, and this restricts the load they are permitted to carry because they must not jam closed in a crash. The structure around the door is therefore designed to carry tensile and shear loads around the door aperture, and of course to pick up the loads, mainly pressure loads, on the door itself. Many doors are designed to fit from the inside, so that the pressure loads act to hold the door closed, and the failure of latching mechanisms becomes less likely and less critical. Large freight doors (Figure 3.8) are

Figure 3.8 Freight door. *Freight doors present particular problems. Their size often dictates that they cannot be opened inwards, so satisfying the need for fail-safety is more complex. A mechanism is incorporated to prevent pressurisation of the aircraft unless the door is closed and fully locked. With doors of this size, the cut-out from the fuselage is so large that the door must carry some of the fuselage loads when closed.* Photograph courtesy Marshall Aerospace Ltd.

normally structural – they carry a proportion of the fuselage loads, reducing the loads carried by the adjacent structure. However, additional structure in the door region will still be required, or stresses will be higher than in a corresponding section without a cut-out. With a large top-hinged freight door, the hoop tension is transmitted through the hinges, through the door and out through latches at the bottom of the door.

The loads arising from large doors may become critical on the ground, when the door is open. In this case the aerodynamic loads arising from the wind on the large surface area of the door generate large forces in the hinges and the door struts.

Thrust loads

With both fuselage- and wing-mounted engines, the thrust will produce a longitudinal force on the fuselage, to overcome the drag of the fuselage. Most engines are fitted with a thrust reversal system for braking, so the direction of the loads can be reversed. The loads are transmitted to the structure through the engine mounts. If an engine fails, the thrust from that engine will be lost, and the thrust from the opposite engine will try to yaw the aircraft. This will need high rudder forces to keep the aircraft flying straight, creating yet more loads on the fuselage. If an engine suffers break-up of a compressor or turbine, it may stop suddenly, and even tear itself free from its mountings. For external engines, the mounting systems may be carefully designed (using shear pins for example) to limit these loads, allowing the engine to break free rather than cause serious damage to the fuselage (or wing) structure if very high loads occur.

Payload loads on floors

The floor of the fuselage acts as a beam, carrying the weight and inertia loads of the passengers, seats, galleys, payload, etc., depending on the type of aircraft and its use. In passenger aircraft, although the total weight being carried by the floor may not be particularly high, very high *localised* loads can occur, especially from small-heeled shoes. Inside the baggage hold of most aircraft, there is a flat floor so that the contents of the hold do not rest directly on the skin. Again, this floor can be subjected to very high local loads when heavy, rigid packages are being loaded, because they can be allowed to fall onto a corner. So the floors of the aircraft need not only to be strong and stiff enough to withstand the overall loads, but need a strong upper surface to withstand high local stresses. The floor is often of composite construction, and in baggage compartments aluminium alloy deck plates or a sandwich of balsa between fibreglass or alloy skins may be used, to provide good resistance to impact damage.

WING LOADS

A wing produces lift by creating unequal pressures on its top and bottom surfaces. The difference in pressure, when multiplied by the area over which it acts, produces the lift force that allows the aircraft to fly. The distributed lift load creates a shear force and a bending moment, both of which are at their highest values at the point where the wing meets the fuselage. During manoeuvres, the lift can be much higher than in level flight, and so these loads are also higher. The structure at this point needs to be very strong, to resist the loads and moments, but also quite stiff to reduce wing bending. The wing will be quite thick at this point, to give the maximum stiffness with minimum weight.

The distribution of the lift loads also affects the size of the bending moment, and so the required thickness at each point across the wing span. The presence of the fuselage reduces the lift of the wing in the area adjacent to it, and some of the pressure near the wing tip spills over from the lower surface onto the top, so the overall distribution is similar to that shown in Figure 3.9. When ailerons or flaps are deflected, the pressures in the area nearby are changed, and the loads will be increased locally. This will further change the shape of the lift distribution (Figure 3.10).

Where engines are mounted on the wings, their weight is obviously going to be borne by the wing structure, along with inertia loads as the aircraft manoeuvres. Thrust forces from the engines will also be carried by the wings. With pod-mounted engines, the thrust force is below the wing, and so this tends to twist the wing. This can be used to balance the effect of the

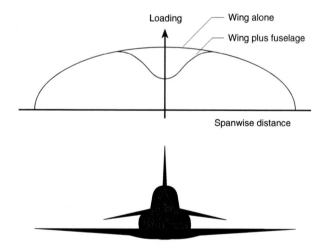

Figure 3.9 Spanwise lift distribution. *With a tapered wing, more lift is produced towards the wing root, and much less lift near the tip. The fuselage also reduces the effectiveness of the wing very close to the root, producing the dip at the centre. Since the bending loads are also much higher towards the root, the wing needs to be thickest near the fuselage and can be quite thin near the tips.*

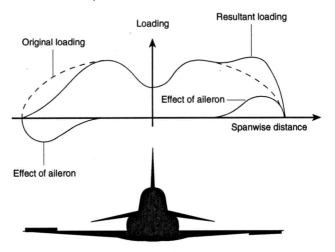

Figure 3.10 Effects of control-surface deflection. *Deflection of the ailerons creates additional loading in the region of the ailerons. This additional loading modifies the overall distribution, and the difference in lift on the left and right sides of the wing causes the aircraft to roll.*

aerodynamics of the wing which create a nose-down pitching moment. Another advantage of wing-mounted engines is that their weight is close to the area in which the lift is produced. This reduces the total fuselage weight, reducing the shear force and bending moment at the wing attachment to the fuselage. (Remember that during flight the *wing* supports the *fuselage*.) So putting the engines on the wings provides *bending relief*.

TAIL LOADS

Like the wing, the tailplane and fin produce lift forces, but these forces need to vary in size and direction to produce the control forces needed to steer the aircraft in pitch and yaw. When the rudder is moved, it creates a force to turn the nose of the aircraft to the left or right. As this force is usually above the centre of the fuselage, it also tries to twist the fuselage. Gusts striking the fin have the same effect. Torsion loads are also created during rolling manoeuvres, by the inertia and aerodynamic effects of the wings, tailplane and fin. Many fuselages are cylindrical in shape, and are able to withstand torsion very effectively.

UNDERCARRIAGE LOADS

Undercarriages create a number of different loads, and all of these must be carried by the structure on which the undercarriage is mounted. When an aircraft lands, high forces, sometimes several times the weight of the aircraft, are created as the main undercarriage absorbs the kinetic energy of the aircraft. At touchdown, the wheels accelerate quickly from rest to match the ground speed, creating a load in the aft direction, and side loads are also generated if the aircraft is sideslipping because of cross winds. The nose-wheel can also experience high shock loads during landing and taxying. During taxying, the loads are smaller, but vary quite widely as the undercarriage absorbs the shocks from any unevenness of the ground. Even during flight, when the undercarriage is retracted, it still has weight and inertia, generating loads on its mountings.

FATIGUE

In every loading case where the direction of the load can change or large changes in the magnitude of the load can occur, fatigue must be considered. Loads can occur in combination on almost every part of the fuselage, and the likelihood of fatigue failure depends on both the applied stress and the number of stress changes to which a component will be subjected. Fatigue is described in detail in Chapter Five.

CHAPTER 4

STRUCTURES

Objectives: to describe the various types of structure and structural members, and understand how airframe structures can be made light, yet strong and stiff, and how the aircraft skin contributes to the overall strength of the structure.

INTRODUCTION

All airframes, whatever the aircraft, are designed using the same principles. The smooth exterior provides a streamlined shape, with extra supporting structure underneath to provide the strength and stiffness needed to operate effectively. In many modern aircraft, the covering and part of the framework are made from a single piece of material. The outer skin, then, hides a complex piece of structure that must be strong, stiff and reliable. In this chapter, the basic principles of structures are described.

The complexity of an aircraft structure can be seen in Figure 4.1, but it can be broken down into groups of fairly simple components, each doing a specific task. The types of component that can be used are described below.

STRUTS, TIES, BEAMS AND WEBS

The structure of most airframe components is made up of four main types of structural member:

1. **Ties** are members subject purely to tension (pulling). Because tension will not cause the tie to buckle, it does not need to be rigid, although it often is. Ties can be made from rigid items, such as tubes, or simply from wire, like the bracing wires on a biplane.
2. **Struts** carry compression loads. Because compressive loads can cause the member to buckle, the design of a strut is less simple than a tie. If overloaded, struts will fail in one of two ways: a long, thin strut will buckle; a short, thick strut will collapse by cracking or crushing, as the material

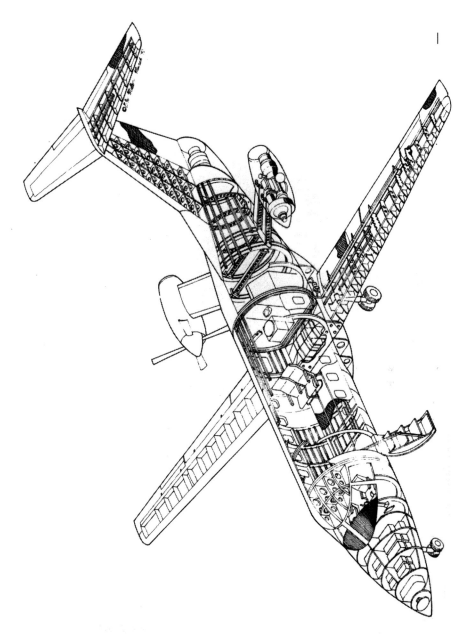

Figure 4.1 Typical aircraft structure. *This drawing is the work of a university student, drawn as part of a design project, and illustrates the tremendous complexity of an aircraft structure. Although the aircraft is a theoretical one, it accurately reflects the structure of many aircraft of its type. Despite its complexity, the structure can be broken down into a number of similar, simpler strutural members, each performing a particular task.* Illustration by Chris Miller.

from which it is made is overstressed. A medium strut may do either, or even both, depending on its dimensions and on other factors. Tubes make excellent struts, because the material is evenly loaded, so that the strength-to-weight ratio is high in compression.

3. **Beams** carry loads at an angle (often at right angles) to their length, and so are loaded primarily in bending. Many of the major parts of an airframe are beams, such as the main spars. The fuselage and wings themselves are structural members, and are beams, because they support the bending loads imposed by weight, inertia and aerodynamic loads.

4. **Webs** are thin sheets carrying shear loads in the plane of the material. Ribs and the skin itself are shear webs. Thin sheets are ideal for carrying shear, especially if they are supported so that they resist buckling.

You may get the impression that each part of an airframe is *either* a tie *or* a strut *or* a beam *or* a web, but this is not so. Some items, such as wing spars, act almost entirely as one type of member, but others act as different members for different loads. For instance, the fuselage skin may be subjected to tensile and shear loads simultaneously. Pure bending loads almost never exist alone; they are almost always related to a shear load. So a beam will normally carry both bending and shear loads.

Figure 4.2 Wing strut. *The strut on the wing of this Grand Caravan supports part of the wing weight and intertia load on the ground and during certain phases of flight, when it is loaded in compression. During normal flight it acts as a tie, transmitting a portion of the lift loads from the wing to the lower fuselage.*

By carefully mixing these members, and making sure that each part of each member is taking its share of the loads, the designer will achieve the greatest strength with minimum weight, and so get the best operating efficiency and maximum safety. It is the designer's aim to ensure that each part of each structure carries a reasonable stress, so that the capability of every part of the structure is used effectively. Only by doing this can the weight of an airframe be made as low as possible, while still providing adequate strength.

There are many uses of struts in an airframe, including the supports for the floor in transport aircraft, undercarriage legs, actuation jacks of all kinds and pushrods for operating flying controls. Struts also frequently act as ties, when the load they take is reversed; again, actuation jacks are typical examples of this.

Consider an aircraft wing made of a solid piece of metal (Figure 4.3). The lift load on the wing can be considered as acting at a single point (the centre of pressure) somewhere near the centre of the wing. The wing is attached to the fuselage at the wing root, and is said to be a *cantilever* beam (attached at one end with no external bracing). The load causes the wing to bend, creating stresses in the material – one side of the beam will be put in *tension* and the other side of the beam will be in *compression*. Along some line, near the centre of the beam, the material will not be subject to either tension or compression, and this line is known as the *neutral axis*.

The wing could be made lighter by removing some or all of the centre. This would have very little effect on its strength or stiffness, because the centre is least affected and therefore it does not carry much of the load. The top and bottom sections must still be strong enough to carry the load (Figure 4.4). You can see that the top section is acting like a strut, and the bottom section acts like a tie.

Of course, it is not possible to remove all of the material from the centre, or the top and bottom surfaces will not stay in place, so some material is needed to keep the correct shape of the wing. What is left, then, is a hollow tubular structure. The top and bottom are called the skins, and take the compressive

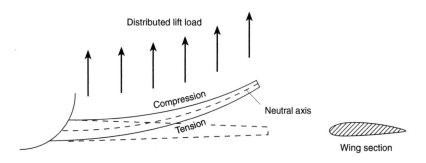

Figure 4.3 Solid cantilever wing. *The diagram shows how a solid cantilever wing would be loaded in flight. The lift load acts upwards on the wing, causing it to bend. This puts the top surface in compression and the bottom surface in tension. Along some line between the two surfaces, the material is neither in compression nor tension, and this line is known as the neutral axis. In practice such a wing would be far too heavy, so some material must be removed whilst maintaining adequate strength and stiffness.*

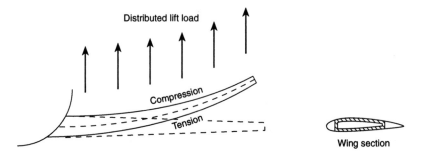

Figure 4.4 Lightened cantilever wing. *Removing material from the centre has little effect on the strength and stiffness, because the material there is under very little stress, but of course it makes an enormous difference to the weight.*

and tensile loads, whilst the front and back faces are webs taking mainly shear loads to keep the shape of the wing. If buckling of the top skin is prevented, by adding extra support, the skin can be made quite thin. The whole structure becomes rather like a rectangular-section tube, and is known as the *wing box*. Leading- and trailing-edge sections are added to the wing box to complete the wing section – these add little to the strength, but give the desired aerodynamic shape. The thin skin will be supported against buckling by *stringers* (Figure 4.5), which take the bending loads created as the skin tries to buckle. Note that they add extra depth to the skin.

Stringers are also used in fuselage for the same reason – if a piece of paper is rolled into a tube, it is very strong in compression. When it does fail, it does so

Figure 4.5 Stringers. *Stringers are used in both fuselages and wings to support the skin, preventing buckling under compression and shear loads by giving the skin extra depth. The stringers shown here are machined as part of the skin.*

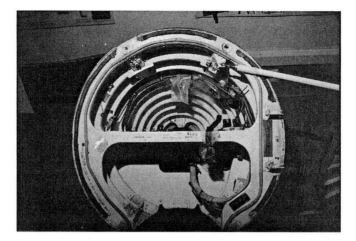

Figure 4.6 Typical beam with built-in ends. *There are many examples of this in structures of all kinds, because they are very efficient at carrying bending loads. The member running transversely across the centre of this composite nose cone acts as a beam, a strut or a tie depending upon the loads applied at any particular time.*

because the wall of the tube buckles in one small area. The tube would be capable of carrying much higher compression loads if the wall were supported. Adding deep stringers would achieve this – the same principle that is used in airframes.

The cantilever structure is widely used in aircraft – they contain many structures that are attached at one end. The wing is just one example of this. Because wings need to be much stronger and stiffer at the root (the attachment to the fuselage), they are wider and deeper there than at the tip, where loads are much less. There are many other examples of cantilever structure.

The cantilever is a special type of beam, because it is supported at one end only. More generally, beams are supported at both ends, and often at other points as well. Supporting beams at more than one point reduces the bending moments, so it is best to support beams as well as possible, since they can be made lighter for a given load-carrying capability.

STRESSED SKIN AND MONOCOQUE CONSTRUCTION

Shear loads occur on many parts of an aircraft, particularly the skin. The whole airframe can be made much stronger and lighter by making the skin contribute to its strength.

In light aircraft the covering may be of fabric, doped to pull it taut and to proof it against air, weather and sunlight, or of thin sheets of wood. The covering is required to withstand the pressure of the air flowing over it, but makes only a small contribution to the strength other than supporting the local loads.

In other light aircraft, and in all large and faster aircraft, sheet metal, normally aluminium alloy, is used. Because of its extra strength, a large part of the load can be borne by this skin, and the underlying structure can be reduced to save weight. This is called *stressed-skin* construction. It also has the advantage of providing a very smooth surface, because the skin is stiff enough not to be distorted by the air flow, which is important for high-speed aircraft. On the wing in particular, it is very important to have an accurate aerofoil shape for best performance.

It is possible to make the skin strong enough to carry all of the load without the need for any supporting framework, a method called *monocoque* construction. In practice, the size of most aircraft would mean that the skin would need to be very thick to avoid buckling, and a compromise, called *semi-monocoque*, is usually used (Figure 4.7). In modern aircraft, it is common for the stressed skin to carry about half of the total load carried by the skin and stringers together. Typical fuselage construction consists of a series of hoops, or *frames*, at intervals along the skin, which give the fuselage its cross-sectional shape, connected by stringers that run the length of the fuselage. Semi-monocoque is an ideal form of construction for fuselages, since it leaves most of the interior free of obstruction, and is used in most aircraft of all sizes.

Figure 4.7 Semi-monocoque fuselage construction. *Stressed-skin construction is most efficient when used in this form, with supporting stringers preventing the skin buckling. Usually, about half of the load is with by the skin, and the other half by the stringers. The fuselage shape is maintained by the frames, leaving most of the inside space free of obstruction.*

INTEGRALLY MACHINED CONSTRUCTION

Joints in aircraft structures, although necessary, can be areas of weakness and can lead to *fatigue* failure, which is explained in Chapter Five. It can be very useful to minimise the number of joints by making large sections from a single piece of material. By controlling the thickness and shape of the material very carefully, the best use of the material strength can be made, and the final airframe will be extremely light and strong. The process involves taking a very large piece of material, usually aluminium alloy, and machining nearly all of it away, leaving a complex component with the material placed exactly where it is needed for maximum strength. Of course, this is very expensive, both in time and metal, and needs large and expensive machines. However, the saving in weight can reduce operating costs of the finished aircraft significantly, and the initial cost may be paid for many times over in service.

Wings for large aircraft, such as the Airbus series, are made in this way (Figure 4.8). It is still not possible to make the wing in a single piece because the machine tools would not be able to reach deep inside the wing. Some joints are also desirable because cracks caused by fatigue will be stopped from spreading further when they reach a joint. This is an important safety feature, and one that must be considered in the design of all aircraft.

Figure 4.8 Integrally machined wing spar. *Many aircraft wings are made in a small number of large and complex parts, each machined from one large piece of metal. This wing spar is for one of the Airbus series. Although very expensive to produce, a large saving in weight can be made, saving many times the extra initial cost in reduced operating costs.* Photograph courtesy British Aerospace Airbus.

CHAPTER 5

MATERIALS AND MANUFACTURING

Objectives: to list the various types of materials commonly used in airframes, some of their applications, and their advantages and disadvantages; to describe some common airframe manufacturing methods.

INTRODUCTION

As has already been stressed, the materials used in structural areas of airframe construction must have adequate strength with minimum weight, in other words a high *strength-to-weight ratio* (SWR). This is not the only consideration, however. Stiffness of the material is often as important as its strength, and other factors need to be considered as well:

- The material must be consistent and predictable in its properties, so that we know what behaviour to expect from it. All materials vary slightly in their basic properties, so it is normal to take the lowest or worst properties, plus an appropriate factor of safety, when using them in design. This gives a reasonable guarantee that the material properties will not be worse than the specified properties.
- It should ideally be homogeneous (having the same properties in all parts and in all directions), although the way a particular material is processed may mean this is not possible. Aluminium alloys are frequently rolled to produce plate and thin sheets, and this means the material properties may be different in different directions. Sheet is assumed to have consistent properties in all directions, but plate does not. If the properties are affected in this way, the final properties must be predictable, and the rolling direction clearly marked on the plate, to leave the material in a useful state.
- Metals must not suffer serious deterioration from corrosion caused by exposure to the weather, sea water or any chemicals that they come into contact with. The effect of stress is likely to accelerate the effects of corrosion. Similarly, non-metals should not be prone to significant degradation under these environments.

- It should be non-flammable or of low flammability (magnesium burns fiercely when exposed to fire, but needs very high temperatures to ignite it). It should present no other safety hazard, such as toxicity, in use, manufacture or repair.
- It should be readily available and at reasonable cost, and should be suitable for manufacturing using standard processes. Where a material's properties are particularly useful, new processes can sometimes be devised to make its use more practical.
- It should not be highly susceptible to *fatigue*, or must be used at stress levels low enough to ensure an acceptable life.
- It must have good stiffness for a given weight.
- It must retain adequate strength at the temperatures to which it will be subjected, particularly with materials used in supersonic aircraft, or in certain regions of the aircraft.

So these requirements limit the types of materials used in airframes, but there are still many options available to the designer. Usually, the particular needs lead directly to one or a small group of materials, but new alloys and new ways of working can change the situation. The following groups of materials meet the requirements listed above, and are used for the main structure of an airframe:

- aluminium and magnesium alloys (*light alloys*)
- steels
- titanium and titanium alloys
- nickel alloys
- plastics and composites

It is difficult to draw exact comparisons between different materials, because there are so many different factors to consider. For example, some resist tension better than others; some resist compression better. Even different types of aluminium alloys are preferred for different types of loads. We can get some idea of how different materials compare by considering their strength-to-weight ratio.

Table 5.1 Comparison of strength-to-weight ratio (SWR) of various aerospace materials

Material	Density	Strength (tensile)	SWR* (relative)
Aluminium alloy (2024-T3)	2.80 g/cm^3	420 N/mm^2	150
Steel (S98)	7.85 g/cm^3	1160 N/mm^2	148
Titanium alloy (Ti-6A-l4V)	4.43 g/cm^3	900 N/mm^2	203
Nickel alloy (MAR-M 246)	7.5 g/cm^3	960 N/mm^2	128
Composites** (carbon/epoxy)	1.40 g/cm^3	920 N/mm^2	657

* appropriate units for SWR are not standardised, so have been omitted here
** composites vary widely in properties, depending on material

ALUMINIUM AND MAGNESIUM ALLOYS (LIGHT ALLOYS)

Pure aluminium and pure magnesium are completely unsuitable as structural materials for airframes, because they have very low strength. However, when alloyed (chemically mixed) with each other or with other metals, their strength is vastly improved, and they form the most widely used group of airframe materials. Alloying metals include zinc, copper, manganese, silicon and lithium, and may be used singly or in combination. There are very many different variations, each having different properties and so suited to different uses. Magnesium alloys are very prone to attack by sea water, and their use in carrier-based aircraft is generally avoided. Aluminium alloys, although denser than magnesium alloys, are much less prone to chemical attack, and are cheaper, so are more widely used. 2024 alloy, known as *duralumin*, consists of 93.5 per cent aluminium, 4.4 per cent copper, 1.5 per cent manganese and 0.6 per cent magnesium, and is the most widely used of all materials in aircraft structures. Aluminium alloys are more prone to corrosion than pure aluminium, so pure aluminium is often rolled onto the surfaces of its alloys to form a protective layer. The process is known as cladding, and sheets of alloy treated like this are known as *clad* sheets or *Al-clad*. Another common means of protecting aluminium alloys is *anodising* – conversion of the surface layer to a form which is more corrosion-resistant by an electro-chemical process. Aluminium–lithium alloys are superior to aluminium–zinc and aluminium–copper alloys in strength and stiffness, so can be used to save weight. Their use is limited because they are around three times as expensive.

An interesting property which certain aluminium alloys share with titanium is that they can be super-plastically formed (SPF). When the material is heated to a certain temperature, far below its melting point, it is capable of being stretched by several times its own length without tearing or local thinning. It can then be deformed, using an inert gas such as argon, to fill a mould and take its shape exactly, with no spring-back when the pressure is released. There are various techniques based on this property, which can be used to make extremely complicated shapes accurately and with minimum weight. The high initial cost of tooling means SPF is limited to certain high-cost items, and it is not yet suited to mass production. Items such as pressure vessels, small tanks and reservoirs may be made using this technique.

Advantages of aluminium and magnesium alloys

- high strength-to-weight ratios
- a wide range of different alloys, to suit a range of different uses
- low density, so greater bulk for same weight means they can be used in a greater thickness than denser materials, and thus are less prone to local buckling; this applies to magnesium alloys even more than aluminium alloys
- available in many standard forms – sheet, plate, tube, bar, extrusions
- aluminium alloys are easy to work after simple heat treatment
- can be super-plastically formed (certain aluminium alloys only)

Disadvantages

- prone to corrosion, so need protective finishes, particularly magnesium alloys
- many alloys have limited strength, especially at elevated temperatures
- magnesium alloys have low strength (but high strength-to-weight ratio)
- no fatigue limit (see section on fatigue later in this chapter)

Table 5.2 Details of aluminium alloys

Alloy designation	Type/composition	Application/description
1000 series	min. 99.0% Al	little use
2000 series	Al–copper	most common type in general use – good fatigue life and fracture toughness
5000 series	Al–magnesium	low density, susceptible to corrosion
6000 series	Al–magnesium–silicon	lower strength than 2000 series, weldable
7000 series	Al–zinc	high strength, but poor fatigue performance
–	Al–lithium	superior strength and stiffness to other Al alloys, lower density, superior fatigue performance; expensive

STEELS

Steel is an alloy of pure iron and carbon (except in stainless steels), with a wide range of other materials. In addition to carbon, steels may contain chromium, nickel and titanium. Steels can be produced with a wide range of properties, ranging from extremely hard and brittle to very soft and *ductile* (able to be bent and stretched). Many steels are very prone to corrosion, including those which have the highest strength. By excluding carbon from the composition, it is possible to produce stainless steel, which does not corrode easily. However, even stainless steels should not be considered totally corrosion-resistant; they may corrode in certain circumstances. Other steels may be protected by plating with another metal, such as zinc or cadmium, although cadmium is used less in modern applications because it is toxic.

All steels share one property – they are dense. Steel finds most usage where its strength can be used to best advantage, for instance where space is limited, or where its hardness and toughness are needed. The most common use is in bolts, shafts and bearing surfaces. It has one more advantage – it performs much better at higher temperature than many other materials.

Advantages of steel

- cheap and readily available
- consistent strength
- wide range of properties available by suitable choice of alloy and heat treatment
- high strength useful where space is limited
- some stainless steels are highly resistant to corrosion
- high-tensile steels have high SWR
- hard surface is resistant to wear
- suitable for use at higher temperatures than light alloys
- most steels easily joined by welding
- very good electrical and magnetic screening
- shows a fatigue limit (see section on fatigue later in this chapter)

Disadvantages

- poor strength-to-weight ratio except high tensile alloys
- dense, so care must be taken not to use very thin sections, or buckling may result
- most steels very prone to corrosion

TITANIUM AND ITS ALLOYS

Titanium and its alloys were little used before the 1950s, but are becoming more widely used now, despite their high cost. The properties are very similar to steel, but they have a superior strength-to-weight ratio. They are widely used in engine components, such as jet pipes and compressor blades, and other components that are subject to high temperatures. Titanium and its alloys can be quite difficult to machine, and suffer from a high degree of spring-back when formed. Many alloys need to be formed at high temperatures, typically over 500°C. Like some aluminium alloys, titanium can be super-plastically formed, allowing very strong and light items, such as pressure vessels, to be made. Titanium also has another related property, that it can undergo diffusion bonding. At elevated temperatures (but far below the melting point), two pieces of titanium forced together under high pressure will fuse and become a single piece. In some ways this is similar to forge welding, but the process takes place at lower temperatures. When combined with super-plastic forming, this allows even greater flexibility of design.

Advantages of titanium and its alloys

- high strength-to-weight ratio
- maintains its strength at high temperatures
- higher melting point and lower thermal expansion than other materials

- can be super-plastically formed and diffusion bonded
- very high resistance to corrosion, especially from salt water

Disadvantages

- expensive
- can be difficult to work, especially machining
- poor electrical and magnetic screening
- very hard scale forms on the surface at high temperatures

NICKEL ALLOYS (NIMONICS)

Nickel-based high-temperature alloys (known as nimonic alloys) are used where very high temperatures (up to 1000°C) will be experienced. For this reason, they find considerable use inside gas-turbine engines, where temperatures are higher than the melting point of many metals. Nickel-based alloys are heavy, and difficult to form, so their use is limited to areas where their properties are essential.

Advantages of nickel alloys

- high strength, maintained up to very high temperatures

Disadvantages

- very dense
- difficult to work

PLASTICS AND COMPOSITES

Pure plastics have little structural use, although it is increasing. However, widespread use is being made of composite structures in aircraft, that is, cloths or tapes of glass, carbon or Kevlar (a trade name for aramid) fibres within a thermosetting resin such as epoxy. Often these materials are made into boards, or composite panels, which consist of a sandwich of, for example, carbon fibre/Kevlar honeycomb/carbon fibre. This makes a panel of limited strength, but which is extremely light, giving a strength-to-weight ratio far higher than a pure metal panel. They are often used for making galleys and bulkheads inside aircraft passenger compartments, but are increasingly used for aircraft structures. Composite panels are not exclusively made from plastics, and aluminium skins or honeycomb cores are commonly used, either together or with plastics (see Figure 5.1). Boron fibres may also be used. The latest generation of fighter aircraft now emerging have up to 30 per cent of the

Figure 5.1 Honeycomb composite panel. *By combining strong outside skins with a core that is resistant to crushing, such as aluminium alloy honeycomb, a sandwich panel is made. This can be extremely light with high strength, and is particularly useful for interior bulkheads, galley structures and leading edges.*

airframe structure made of composite materials. The Lear Fan 2100 business aircraft has a structure that consists of 77 per cent composite materials.

The fact that many composites are based on fibres means that the strength and stiffness are not the same in all directions. This is not always a disadvantage, since many structures are loaded primarily in one direction. By laying up the fibres mainly in that direction, the best use of the material's properties can be made. In this way, the structure can be tailored to its usage in the airframe.

Many plastic-based composites show lower tolerance of impact damage than metals – for instance from bird strikes. Some composites can be quite difficult to repair safely. Kevlar, for example, absorbs water if damaged, which can make it difficult to make a satisfactory repair. In all cases, very careful repairs are needed, usually requiring heater mats and vacuum pumps. It is often easier to replace a damaged item, returning it to the factory or repair facility or discarding it. In some cases, this is not convenient, for instance a battle-damaged aircraft needing to be returned to service as soon as possible. In any event, this method usually means that the user needs more spares, which often means higher costs.

Another problem with the increasing use of composites, in leading edges for example, is that they do not provide electrical or magnetic shielding for cables. High voltages and currents can be induced in the aircraft's electrical system, which may cause it to fail, if the aircraft is operating close to strong electromagnetic fields, for instance on the deck of a ship. This may also occur if its own radar is close to the equipment concerned, or if a nuclear explosion occurs even many miles away. If the aircraft structure cannot provide enough protection, extra shielding is needed, which adds weight, cost and complexity.

However, with all of these reservations, composites used carefully can produce great weight savings, and aircraft contain increasing amounts of composite structures. Within decades, they may well replace aluminium alloys as the primary material used in airframe construction.

Advantages of composites

- very high strength-to-weight ratio and low weight (varies between materials)
- non-corrodible (but some materials absorb water if damaged)
- easily available in a wide range of forms
- can make complex shapes easily
- directional nature of fibres can be used to produce optimum strength in direction of highest loading
- low resistance to radar and radio signals is ideal for radomes and antenna covers

Disadvantages

- need special manufacturing, inspection and repair methods
- some materials prone to impact damage
- strength and stiffness not the same in all directions
- poor electrical screening

FATIGUE

Fatigue is a material's tendency to fail under a large number of repeatedly applied stresses, such as those created by gusts, turbulence or vibration. Stresses may be much lower than that which would cause failure in a single application, but may still create a fatigue failure if enough cycles are applied. As the applied stress increases, the number of applications before failure occurs will progressively reduce. Next to human factors, fatigue is the main cause of aircraft accidents. Some materials, although not as good as others in terms of strength-to-weight ratio, show superior fatigue properties, and are extensively used for this reason. The fatigue properties of a material are often shown by a family of *S–n curves*, which show applied stress against the number of cycles to failure (e.g. Figure 5.2). There is a family or set of curves for each material because the curves are dependent on the state of heat treatment, the mean stress and other factors.

Fatigue performance is highly dependent on the material selected, and it is common to use different alloys in differently stressed areas.

Steels are unusual in that they show a fatigue limit – a stress below which fatigue damage does not occur – and the life of the material becomes infinite. Many other metals do not show this property.

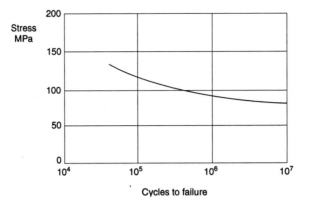

Figure 5.2 S–n curve for aluminium 2024-T81 alloy. *The normal way of showing a material's fatigue properties is the S–n curve, which shows the applied stress against the number of cycles or applications of stress required to cause failure. The S–n curve for each material or alloy is different, and also varies with loading conditions and state of heat treatment. Note that the horizontal axis on this graph is logarithmic, with each interval representing a ten-fold increase over the previous one.*

The area of the fuselage that is most critical in fatigue is the top skin, because tensile stresses from bending and pressurisation loads are additive. The wing-to-fuselage joint is also a problem area, because the load transferred from the wing spars causes bending of the fuselage frames, and can generate fatigue cracking in the frames close to the joint. The likelihood of fatigue failure in all areas can be reduced by avoiding stress concentrations such as sharp corners and closely spaced holes.

To guard against fatigue failures, aircraft life used to be quoted in flying hours, although modern aircraft use a system of fatigue monitoring and fail-safety – the provision of multiple load paths, so that an alternative load-carrying member can carry the full load if a section of structure fails. Fatigue meters are fitted to many aircraft; the amount of fatigue life consumed during each flight can be assessed by monitoring aircraft accelerations against time. The manufacturer carries out a long series of fatigue tests on an airframe, to identify in advance where problems are likely to arise. Comparing the information recorded by the fatigue meter with the manufacturer's tests allows remaining airframe life to be constantly monitored to guard against unexpected fatigue failure.

Another possibility for guarding against fatigue failure is by using fatigue fuses. These are special devices that are bonded to a piece of structure, and are designed to break through the action of fatigue some time before the structure does, giving advance warning of problems. They are monitored electrically or by regular inspection.

MANUFACTURING METHODS

There is a wide variety of manufacturing methods available to the designer, and each is particularly suited to producing specific types of structure, and often dictates or depends on the material selected. Most manufacture can be broken down into two main types of process – material forming and joining – although not every manufactured item involves both types of process. In some forms of manufacturing, both processes are combined; an example can be found in composites manufacture.

Material forming

Metals can be formed by a number of processes:
- bending, pressing, rolling and drawing
- machining
- forging
- casting
- extrusion

Bending, pressing, rolling and drawing

Although very basic techniques, these are highly versatile. Bending can normally be performed using simple tools, but components may also be made using a *press*, which squeezes or stretches a piece of material (or blank) between two shaped dies until the shape of the material conforms to that of the dies. This technique allows quite complex components to be made in one or several operations. Fuselage and wing stringers (described in Chapters Four and Six) can be made by rolling – passing a strip of alloy through a set of rollers that progressively bend the strip into a Z, J or top-hat section. Once correctly set up, the resulting components can be very accurate, and consistent along their length. Because the entire strip is not bent at once, there is theoretically no limit on the length of each section made. In practice this is limited by the length of material and the working space available. Adjusting the settings of the rollers allows curved bends to be produced, or the strip to be curved. This is ideally suited to producing fuselage frames (Chapter Eight). A similar technique may be used to make tubes – a strip of material is passed between rollers that progressively curve the strip across its width until the sides meet. The sides can then be joined to make a rigid tube, for example by welding.

More simple bends can be made by folding bars (Figure 5.3), which have a pair of straight jaws that clamp a sheet firmly in position, and another jaw that can be swung upwards to create the bend. Whatever the method used, it is important that very sharp bends are avoided, because the difference in length between the outside and inside faces of the bend must be accommodated by stretching and compression of the material. If the bend radius is too small, the material will crack, or at least suffer a considerable reduction in strength.

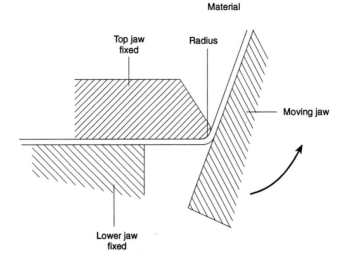

Figure 5.3 Folding bars. *A fixed pair of jaws holds the material firmly in position whilst another jaw is pulled upwards to create the bend. The top jaw has a radius selected to match the thickness of the material – too small a radius will cause the material to crack.*

Machining

This describes the action of removing material from the initial *billet* (a block of material from which the component is to be made), or a plate, bar or other form, by moving the material against a static cutting edge, by moving cutting edges against the static material, or both. *Turning*, using a lathe, involves rotating the billet and bringing a static cutter against the moving surface (Figure 5.4). Since the billet is rotating, the resulting cut will be at a constant radius about the axis of the billet, so turning is used to produce shafts, tapers and cylinders. Using a suitably shaped cutter and passing it down the length of the rotating billet can create a screw thread.

If the billet is clamped firmly and the cutter is rotated, the process is known as *milling* (Figure 5.5), and can produce both flat and curved surfaces. Using shaped cutters allows more complex cuts to be produced, and in some cases a cutter will be specifically produced for a certain task.

A *shaping* machine uses a cutter that moves in a straight line across a fixed billet, a little like a wood plane. A variation on this technique is broaching, which has a long tool (the broach) carrying a large number of shaped cutting edges, each of which takes off a little more material than the previous edge. The final cut will conform to the profile of the final cutting edge of the broach. Broaches may be several metres long. The process is used for cutting the complete root attachment slots in turbine discs, to receive the individual turbine blades.

Other machining methods include drilling, grinding and spinning.

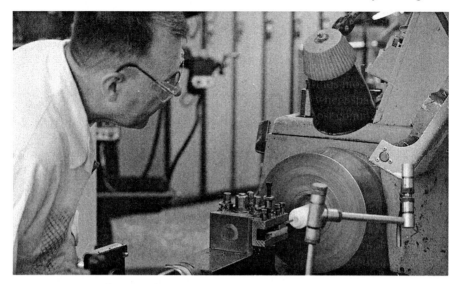

Figure 5.4 Turning. *A lathe is used for the process of turning, where the material to be shaped is rotated and a static cutter removes material in a circular cut. This process is used to make shafts, tapers and cylinders. If a suitably shaped cutter is moved along the shaft during cutting, a screw thread can be produced.*

Figure 5.5 Milling. *Milling differs from turning in that the material is stationary and a rotating cutter removes material. This allows flat surfaces and non-circular shapes to be produced.*

Forging

Forging is a process of forming a billet of material to the required shape by compressing the billet between shaped dies, usually at high temperature. The material is forced to take up the shape of the dies by the very high pressures involved, as the dies are brought together by hammer-like blows. It can be used to produce strong components in a variety of metals, for instance undercarriage components, but does not give the precise control of dimensions that is associated with machining.

Casting

Casting can be used to create highly complex shapes, but machining may be required to achieve the required finish, particularly on surfaces that need to be accurately located, or to incorporate holes or threads. The material to be cast is melted in a furnace, and the molten metal is poured into a mould, which has a cavity the same shape as the required product. The material fills this cavity, and solidifies as it cools. Apart from simple shapes, the mould will usually need to be destroyed to remove the component. The mould is produced by a pattern, which is identical in shape to the finished casting. The mould may be made of sand or a plaster of Paris type material (sand casting), or it may be a metal die (die casting). It is obviously impractical to destroy a metal die to remove the component, so this places some limitations on the shape of die-cast components. Cast components of highly complex shape may be made using a wax pattern, which can be assembled from a number of simpler parts. As the material is cast, the wax melts and runs out of the mould, leaving an accurate mould cavity. Lost-wax casting, as this process is called, makes it possible to produce components with re-entrant cavities and other features that would not be possible if it were necessary to remove the pattern before casting.

Extrusion

Extrusion is the process of pushing material through a hole of a defined shape to produce a continuous length of material of a given section (Figure 5.6). The 'hole' is called a die, and the extruded part adopts a section identical to the shape of the die. The process may be carried out at room temperature, or some degree of heating may be employed, but the material is not molten. The process can be seen simply by squeezing a tube of toothpaste – the paste extrudes through the neck of the tube, adopting the size and shape of the hole. Small cross-section extrusions may be rolled onto a drum as they are produced, but larger ones will be drawn as straight lengths, and cut to length during production. The cross-section of the extrusion is determined entirely by the die, and ridges, grooves and even tubes may be incorporated. The cross-section will of course be uniform along the length of the extrusion, and any variation of this cross-section is limited to removal of material by machining.

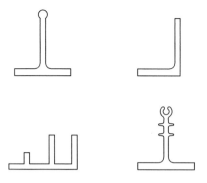

Figure 5.6 Typical extruded sections. *These sections are typical of the complex shapes that are possible using extrusion. The material takes the shape of the die through which it is passed, and since the shape cannot be altered during the extrusion process, the extrusion must have a constant cross-section along its length.*

Super-plastic forming (SPF)

This may be performed on certain aluminium and titanium alloys, and on pure titanium. At a temperature of about half the melting point, the material becomes capable of undergoing a high degree of extension without local thinning, provided the rate of extension is correctly controlled. The material may be stretched by 300 per cent, or up to 1000 per cent for some materials. The usual method for this process is to clamp a sheet of material inside a mould, and raise the temperature to the correct process value. An inert gas such as argon is introduced to one side of the sheet at high pressure, and the metal expands into the mould (Figure 5.7). The rate of expansion is carefully controlled by controlling the gas pressure. Alternatively, a pair of shaped dies may be used to press the material into shape. When the forming is complete, the component is cooled, and the mould is opened to reveal a part that accurately reproduces the shape of the mould. A hard scale which forms on the surface during forming is then removed chemically, and the component is complete. The benefits of this technique are the accuracy of forming, plus the complexity of shape that can be produced. SPF components are generally much lighter than fabricated alternatives, for those components that lend themselves to manufacturing by this method.

Composites

These are generally formed to the required shape using a mould. Most aerospace applications use non-metallic fibres in an epoxy base. The fibres are available in a range of forms, from tissue and chopped-strand mat to woven cloth, tape and spun fibre. Most are available with the resin already applied (pre-pregnated), and will be laid up in the mould to the required thickness. Because the properties of the finished product will be highly dependent on the

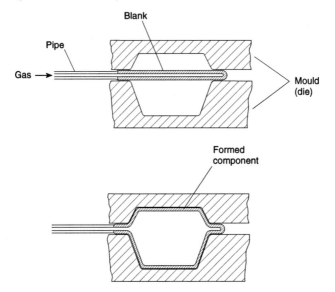

Figure 5.7 Super-plastic forming. *When certain suitable materials are heated to a specific temperature, well below their melting point, they can be stretched by several times their normal length without tearing or local thinning. Sheets or other simple forms of the material are heated in a press, and dies or gas force the material to take up the required shape. When the material is removed, it conforms accurately to the shape of the mould.*

direction of the fibres, this is carefully planned during the lay-up stage. When the lay-up is complete, the component is cured using heat, and pressure is applied to force excess resin from the mould. Pressure vessels may be created using spun fibres wrapped around a mould (filament winding).

Grain

One consideration when determining the process to be used is the effect on the grain pattern of the material. The raw material will have a grain structure that is essentially the same in all directions, but any deformation of the material will deform the grain structure accordingly. Forged components generally have the grains distorted so that they follow the outline of the component, which gives high strength. Where the properties are important, then rolling direction must be taken into account when the material is used. Machining does not affect the grain structure, since it removes rather than deforms material. For this reason, rolled threads are slightly stronger than machined threads, and are preferred in many applications. Casting, like machining, produces a uniform grain structure, although the size and atomic distribution within the grains may be a function of cooling rate.

Joining materials

There is a wide range of methods of joining materials, some of which are suited to general use and some which are specific to certain materials or material forms. Common methods are:

- riveting
- bolting
- bonding
- welding

Riveting

This is the most common method of joining sheet materials in aircraft. A hole is drilled through the sections to be joined. A suitably sized rivet is put into the hole and set – the tail of the rivet is deformed so that it expands sideways and grips the sides of the hole, clamping the material together. They may be set by hand, but more often a powered riveter is used. Rivets are available with a variety of head shapes, including mushroom, snap and countersunk heads. Where access is available to only one side, blind rivets are employed (Figure 5.8). These are hollow rivets, which are set by drawing a mandrel through the hollow rivet from the same side as the head; the mandrel then breaks off. Because blind rivets are hollow, they must generally be sealed separately if

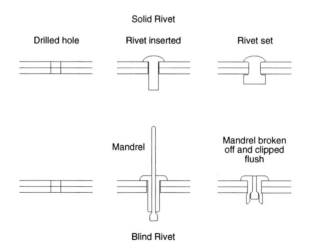

Figure 5.8 Solid rivet and blind rivet. *The rivet is pushed into the drilled hole, then set – the tail is deformed so that it expands sideways. The deformed tail clamps the sheets together and retains the rivet in the hole. The blind rivet may be used when access is restricted to one side. Pulling the mandrel through the rivet expands the rivet to set it, and the mandrel breaks off to leave a neat fastener. Blind rivets are hollow, and need to be plugged after setting if sealing is required.*

required. There is a wide variety of special rivets, each for a specific application. Rivets may be made from any deformable material, but light alloy rivets are almost universal. An exception is the high-shear rivet, which uses a steel stem for very high shear strength, with an aluminium alloy collar that is deformed during setting to hold the rivet in place. All rivets are designed to be used in shear, and have limited strength in tension.

Bolting

This is most useful where high shear loads or significant tensile loads will occur at the joint. Aluminium bolts are available for shear applications, but generally steel bolts are preferred for most applications. In most airframe applications, the fasteners must be locked to ensure that the nut and bolt do not loosen under vibration or temperature changes. A variety of methods can be used, including split pins, wire locking and clinch nuts. For bolts fitted to blind holes, wire locking is the most common method.

Bonding

Bonding is the technique of joining materials using special adhesives, generally of the Redux type. Redux is a hot-melt, hot-cure adhesive, which is available in sheets. The sheet is cut to size and the backing paper is peeled away. The adhesive is then placed between the materials to be joined. The joint is clamped securely and the assembly cured in an oven. The adhesive melts and flows to fill the gap, then cures to form a strong bond. Although it is not as strong as riveting, the joint area is much greater, and it is comparable in many applications provided the temperature is kept within the working limits. Its peel strength is limited, but this may be overcome by bolting or riveting at each end of the bond. The advantage of bonding is that it involves no drilling, so sealing of wing boxes used as fuel tanks is made easier.

Welding

Welding (Figure 5.9) involves melting the two metals to be joined so that they fuse to become a single piece (fusion welding) or heating the material close to its melting point and joining the two surfaces under pressure (forge welding). With fusion welding, filler metal is added where required to fill gaps. Heating of the material to be fusion welded may be achieved using a gas flame (typically acetylene and oxygen), an electric arc or by a high-energy beam of electrons (electron-beam welding). The latter allows great accuracy in the welding process, producing a very small pool of molten material. Forge welding is usually performed by passing an electric current through the material to heat it, and the pressure is also applied through the electrodes. Spot welding, used commonly in the assembly of car bodies, is an example of this, as is seam welding. Welding can be performed on many materials, including light

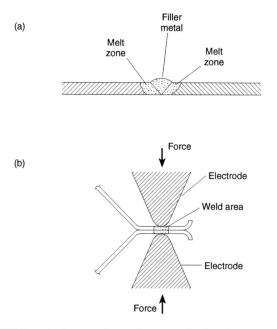

Figure 5.9 Welding. *In fusion welding, the material is locally melted on either side of the required joint, so that the edges of the weld fuse together. Extra material, the filler material, is normally added. Forge welding, for example spot welding, does not involve melting – the material is heated close to its melting point and joined under pressure, so that it fuses.*

alloys, although it is most common on steels and Nimonics. It is important that the weld is uniform and contains no inclusions – particles of unwanted material or gas pockets – or the strength will be greatly reduced. Inspection of welded joints is not easy without cutting through to examine them, but X-ray techniques, although expensive, can be used for critical applications.

The related techniques of brazing and soldering take place at much lower temperatures than welding – only the filler metal is melted so that it flows into the gap. The filler then forms an alloy at the boundary, creating a joint with less strength than welding but adequate for many applications. In all cases, the temperatures involved may change the state of the materials to be joined, and further heat treatment is likely to be necessary.

Diffusion bonding

This is often associated with super-plastic forming of titanium and its alloys, since the two processes are often used together. At similar temperatures to those used for SPF (880–920°C), two pieces of titanium that are clamped accurately together fuse to form a single piece. This appears to be similar to welding, but the temperatures involved are far lower than the melting point, and complete fusion occurs. If a properly bonded joint is cut, there is no visible joint line – the two pieces fuse completely, leaving no reduction in material properties. When combined with SPF, the process is known as SPFDB ('spuffdubb'), and can produce complex parts with high strength and low weight. Like welding, it is difficult to inspect the bond without cutting, and process control is critical. In areas where bonding is not required, a stop-off compound is placed between the two layers to prevent contact. Diffusion bonding is commonly used in the manufacture of turbine blades for gas turbine engines.

Composites

Because of their limited properties, composites require special techniques for joining them. The strength of any joint is limited by the material strength, and methods that involve small contact areas, such as riveting, are not generally suitable. Honeycomb sandwich panels, a common use of composites, may be joined by fitting an insert into the panel using a resin which is then cured. The resin fills a cavity created inside the panel, spreading the load over a larger area. The insert is threaded, allowing a bolt or screw to be fitted. Other methods operate on similar principles, for example by incorporating a metal plate that is drilled for attachments, or bushes to prevent bolts crushing the material. Special rivets and other fasteners are also available, all having the characteristic of spreading the load over a reasonable area. For some joints, woodworking techniques such as dovetail joints can be used.

Heat treatment

Many of the processes employed in forming and joining materials involve heating, and this may change the properties of metals. It is common for heat treatment to be carried out after any process involving heating. This returns the material to the desired state, and ensures the correct strength and fatigue properties.

It is common for alloys to be produced with widely varying properties, purely as a result of different states of heat treatment.

CHAPTER 6

WINGS

Objectives: to describe the basic components of typical wings, various manufacturing methods and some specific features of modern wings.

INTRODUCTION

As already described, different sizes and types of aircraft need different construction. This applies to the mainplanes, or wings, as much as to any other part. Each wing is basically made up of two parts – *the internal structure*, such as spars and ribs, and the *skin*, which can be of fabric, metal or composites – although the distinction between structure and skin may not be readily apparent in modern fast jets or large transport aircraft.

SPARS, RIBS, STRINGERS AND SKIN

Wings are made up of a large number of components, even with integrally machined structures, but the structural part consists of four main types of component:

Spars

Most of the lift, and hence shear force, that occurs on the wing is collected together into the spars. Spars run spanwise, in other words from the root (where the wing is attached to the fuselage) to the tip. Most wings contain two spars – the front and rear spars – but it is quite common for wings to have more than this. This is particularly so with swept wings on transport aircraft, which often have a short spar, the *auxiliary spar* or *kick spar*, which helps to support the undercarriage, and provides a location for the inboard flaps.

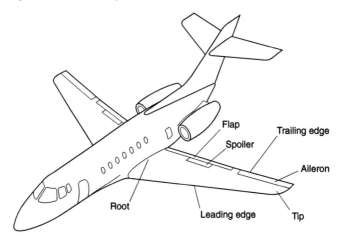

Figure 6.1 Parts of the wing – external.

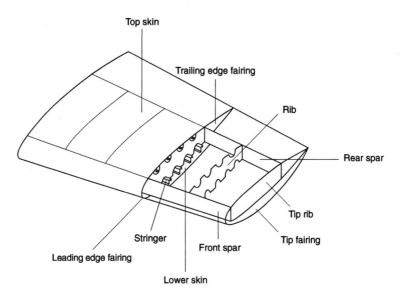

Figure 6.2 Parts of the wing – internal.

Ribs

Ribs give the shape to the wing section, support the skin and act as baffles to prevent the fuel surging around as the aircraft manoeuvres. They collect loads from the control surfaces, undercarriage and stores, where fitted, and pass them to the wing skins and spars. They also help to stiffen the wing structure, particularly against twisting. Ribs normally run either fore and aft or at right angles to the spars, as shown in Figure 6.3.

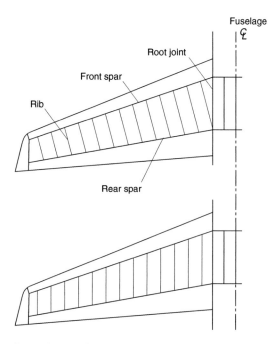

Figure 6.3 Wing ribs. *Ribs give the wing its aerofoil shape, support the skin to prevent buckling, distribute the loads from flaps, ailerons, etc. and also act as baffles to restrict fuel movement where the wing box also makes up the walls of the fuel tanks. The two examples show ribs taking the shortest distance between the spars, and in the direction of the airflow. The former is more common in larger aircraft, and the latter is preferred where the skin is of a flexible material.*

Stringers

Stringers (Figure 6.4) are attached to the wing skin, and run spanwise. Their job is to stiffen the skin so that it does not buckle significantly when subjected to compression loads as the wing bends and twists, and loads from the aerodynamic effects of lift and control-surface movement. To be effective, they must be attached more or less continuously along their length, either with closely spaced rivets or by *bonding* (gluing) along their entire length. Advantages of bonding are that there are fewer holes through the skin which would need to be sealed to prevent fuel leakage, cost of manufacture may be reduced and fatigue properties are good.

Skin

Most wings are of *stressed-skin* construction (see below), so the skin performs several tasks. It gives the wing its aerodynamic shape, it carries a share of the loads caused by the pressure changes around the wing in flight and it helps to carry the torsional loads generated by features attached to the wings – ailerons,

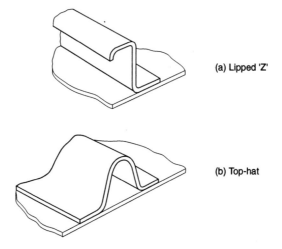

(a) Lipped 'Z'

(b) Top-hat

Figure 6.4 Wing stringers. *Stringers stiffen the skin, preventing buckling under torsional, shear and compressive loads. They also carry part of the load when the skin is in tension. They can be made in a variety of sections, the most common being the Z and top-hat sections. The lip on the Z stringer stiffens the flange, stabilising the stringer in compression.*

flaps, wing-mounted engines. As most aircraft carry fuel in the wings, it also acts as the walls to the fuel tanks, together with the front and rear spars. To allow inspection and maintenance, it will contain a number of small inspection covers, which can be removed for access to the wing structure.

STRESSED-SKIN WINGS

Because air loads increase as the square of the speed, at the speeds associated with most types of flight the skin needs to be very strong to support the loads it will experience. Aluminium alloys are the most often used, and they are so strong that they could be used in very thin sections. However, a thin skin is flexible, and relies on the supporting structure to provide the required stiffness. If the skin is a reasonable thickness, perhaps of the order of one millimetre, it can share the loads taken by the structure underneath, which can then be made lighter. Almost all aircraft have their structure made entirely in metal, or a mixture of metal and composite materials.

- The two main spars are the main structural members, but a large contribution to the strength and stiffness is made by the skin and stringers.
- The entire wing box is normally in metal construction, although the wing tip, ailerons and leading edge may be of composites. To reduce weight the ribs often have large lightening holes, with flanged edges to maintain the required stiffness.

- The skin may be fixed to the internal structure by rivets or by bonding, using special adhesives such as *Redux* hot-cured adhesive.
- The volume between the front and rear spars is often used for storing fuel, and holes in the ribs allow a controlled flow of fuel inside this space. The leading- and trailing-edge sections are used for carrying electrical cables, control wires, cables and push-rods, hydraulic pipes and other items along the wing, and the leading edge may also carry anti-icing pipes.

Using the skin to carry part of the loads in this way is called *stressed skin*, and it allows thin cantilever wings to be produced which are strong enough to resist the loads generated at high speeds. Stressed-skin construction is the only viable option for aircraft of medium to high speeds.

Spar design

A spar must have some depth so it may resist the bending loads imposed on it. An example of this is an ordinary ruler, which will flex easily when loaded on its top or bottom surfaces, but is very stiff when a load is applied to the edge (Figure 6.5).

The spar has the majority of the material in the top and bottom sections, called the *caps*, with a relatively thin web between them. The caps may be separate pieces, or they may include half of the web – see Figure 6.6. This creates a *crack stopper*, which is simply a joint that will act as a barrier if fatigue cracks begin to appear in the spar. In a two-part spar, only one part will break, and the spar is designed so that the other part will be strong enough to

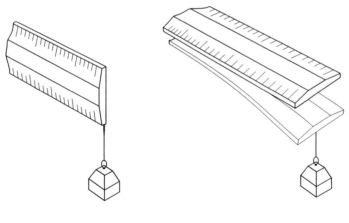

Small deflection   Large deflection

Figure 6.5 Ruler used as a spar. *An ordinary ruler is very flexible when loaded across its thickness, but is very stiff when loaded across its width. The same applies to any beam, including wing spars, because stiffness is dependent on the depth in the direction of the load.*

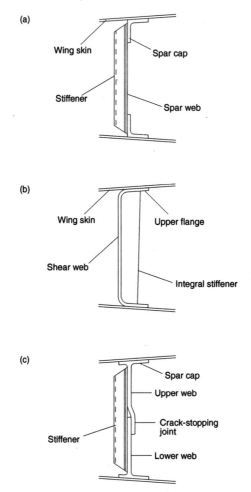

Figure 6.6 Typical spar sections. *Spars may be produced in a variety of shapes, but what they have in common is that most of the material is concentrated in the top and bottom caps or flanges, where loads are highest. A relatively thin web between them obliges the caps to operate in unison as a single beam. The webs essentially carry shear forces. The fail-safe joint is effective because any cracks will not progress beyond the joint.*

carry the entire load. This is known as *fail-safety*, and is an important part of most aircraft design. If the spars are made in three parts, then any two parts must be able to carry the loads, so that any one part may fail without the possibility of the aircraft structure failing entirely. The failed component will therefore not cause a hazard before it is rectified during the next inspection.

A single-spar design is not likely to be particularly stiff in torsion, so the wing may twist excessively in flight. Most modern aircraft use two main spars, with stressed skin between them, to form a *torsion-box* construction (Figure 6.7). The leading and trailing edge sections are then added in a lighter construction, often composites, and make a negligible contribution to the overall strength of the wing. The major advantage of this arrangement is that the space within the torsion box is an ideal space to store fuel. The whole volume is treated with sealants to prevent leakage, and is divided into several large tanks. The fuel may then be moved around as required to balance the aircraft or reduce loads in flight. The skin must be stiffened to prevent buckling, and stiffening stringers will be bonded or riveted to it, or they can be *integrally machined* as part of the skin.

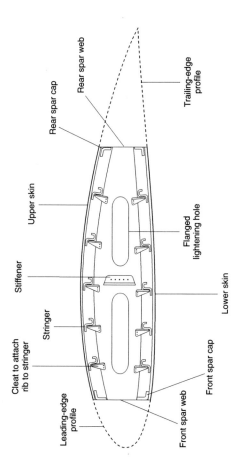

Figure 6.7 Two-spar wing torsion box. *The two spars form a torsion box with the top and bottom skins. This is a very stiff, light structure, with a useful space for storing fuel. The leading and trailing edges do not carry any significant loads, but are there to complete the aerodynamic shape and act as covers for cables, control rods, etc.*

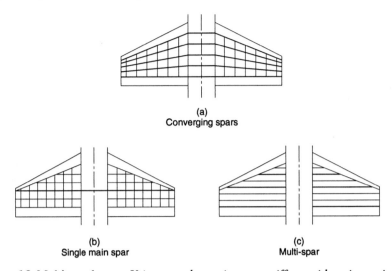

Figure 6.8 Multi-spar layout. *Using several spars improves stiffness without increasing thickness. Such a wing is ideally suited to very-high-speed flight. It also increases fail-safety and tolerance to battle damage.*

On some high-speed aircraft, a multi-spar wing is used (Figure 6.8). Using a larger number of spars allows a very stiff wing to be produced without increasing thickness, so is highly suited to high-speed flight. This construction would normally be made using integrally machined spars, with the wing constructed from a small number of large-scale machinings, as described below.

MACHINED SKIN

As an alternative to making stressed skins by attaching stringers to the skin (*fabricated*), the skin, stringers and spar flanges can all be machined from a single piece of alloy, called a *billet*. This billet may be many metres long, since it is possible to make the top or bottom skin for one wing in a single piece. The billet is much thicker and heavier than the final machining, with perhaps 90 per cent of it being removed during machining. Although this is very expensive, in both material and machining cost, the final result is a lighter and stronger skin than a fabricated one.

Advantages of machined skin

- riveting is not required, so a smoother surface is achieved
- lighter structure with more even loading than a fabricated one
- computer-controlled machining means mistakes or faults are less likely, and more easily detected

Figure 6.9 Machined skin. *By machining the skin and some local reinforcement from a single piece of material, called a billet, a large saving in weight can be achieved, since all unnecessary material can be removed. The number of joints is reduced, reducing sealing problems. However, this method is more expensive than fabricating the skin from sheet materials, and the extra cost must be covered by savings in operating costs. Close attention must be paid to proper crack-stopping features, to prevent the unrestricted spreading of cracks during the life of the aircraft.* Photograph courtesy British Aerospace Airbus.

- easy inspection during manufacture and in service
- no maintenance required
- easy sealing of fuel spaces

Disadvantages

- high cost, particularly setting up tooling
- for combat aircraft, battle damage repair can be more difficult
- careful design needed to maintain fail safety by limiting spread of fatigue cracking

Because they are very different in shape to other types of wing, delta and heavily swept wings have different construction. Delta wings have a large chord at the wing root, and so providing adequate thickness for structural stiffness is not a problem. Swept wings may be required to house the undercarriage when it is retracted, and with swept wings it must be located near to the trailing edge. As previously described, a solution to this is to add an auxiliary spar, and to increase the chord of the wing at the root. This then gives enough depth in the wing to accommodate the retracted undercarriage, and provides a strong point for the undercarriage mounting.

HARDPOINTS

Ideally, combat aircraft should be capable of carrying a wide range of weapons, with the facility to change weapon types quickly for different missions. This is achieved by locating *hardpoints* at various places on the aircraft, normally below the wings and fuselage (Figure 6.10). The hardpoints on the wings are called *pylons*, and each pylon will be capable of carrying a range of stores, up to the maximum weight allowed for that pylon. The mounting arrangements of the stores to the hardpoints are standardised to allow for this. The inboard pylons, those nearest to the fuselage, are usually *wet* pylons – they are capable of transferring fuel from special under-wing tanks. This allows the range or duration of the aircraft to be increased, which can be used to advantage. In an emergency or for combat, the weight and drag penalty of the tanks can be eliminated by jettisoning them, allowing them to fall away from the pylon. Other stores, such as missiles, need special wiring to allow them to be used, and

Figure 6.10 Hardpoints. *By using standardised hardpoints, a range of external stores can be quickly and easily fitted to several different aircraft types without modification. Fuel tanks need 'wet' pylons (equipped to transfer fuel), normally only the inboard ones.*
Photograph: Alistair Copeland.

certain hardpoints may be wired to allow a range of such stores to be fitted. Missiles need a launcher from which to be fired; the launcher has the standard fittings to allow it to be fitted to several aircraft without modification.

FUEL STORAGE

As already described, an aircraft needs a large amount of fuel to be capable of operation over a useful distance. In commercial aircraft this is usually around a quarter of the aircraft's maximum operating weight. Such a large amount of fuel is difficult to store, since it takes up a large volume. The usual solution is to store most or all of the fuel inside the wing. Because of the long, thin shape of this volume, the wing is divided into several tanks, each one usually having its own pumps. This allows fuel to be moved between tanks in flight, which changes the trim of the aircraft to minimise drag. The fuel may be contained in rubber bags, shaped to fill the available space, or the inside of the wing structure may be sealed using special rubber compounds to prevent leakage.

Combat aircraft often have quite small wings, so the volume inside them may be too small to carry a reasonable quantity of fuel. In this case, extra tanks may be fitted in the fuselage, again usually of the bag type, which are self-sealing in the event of damage. To extend the range further without losing performance, jettisonable external fuel tanks may be fitted to wing pylons.

UNDERCARRIAGE MOUNTS

The position of the undercarriage must be set in relation to the aircraft's centre of gravity. To maintain correct stability, the position of the wing must also be related to the centre of gravity. For most commercial aircraft, this means that the undercarriage needs to be positioned near the trailing edge of the wing. The main undercarriage legs need to be a reasonable distance apart to make the aircraft stable during take-off and landing. It is convenient to attach them to the wing, a short distance from the root. The undercarriage is normally retracted for low drag, and it usually folds into an *undercarriage well* (Figure 6.11), in the wing and/or the lower fuselage.

The undercarriage and its mountings are subjected to very high forces on landing, and these loads must be safely transmitted into the wing, so the structure around the undercarriage mountings is reinforced. When retracted, doors close the well, to provide a low-drag arrangement.

FLAPS, SLATS, SPOILERS AND LIFT DUMPERS

Most aircraft need to land as slowly as possible, for safety. Providing enough lift to support the aircraft in flight at very low speeds would require a design that

Figure 6.11 Undercarriage and well. *The undercarriage is subjected to very high loads during landing, and must be strong, as must its mountings. Its position is normally fixed by other factors, but can often be accommodated inside the wing. The undercarriage well is closed by doors after retraction, to minimise drag.*

would not be efficient at the higher speeds at which most aircraft cruise. So extra devices are added to the leading and trailing edges of most aircraft to increase lift when deployed. Other devices may be deployed to reduce the lift produced by the wing, to allow a steeper approach or provide a positive touchdown.

Flaps and slats

Flaps are fitted at the trailing edges, and attached to the rear spar and/or the auxiliary spar. There are several different types of flap, of varying degrees of complexity and effectiveness. Light aircraft will usually have simple flaps, if any. Larger aircraft have the more complex split flap or Fowler flap. Most large transport aircraft have double-slotted Fowler flaps, as shown in Figure 6.12. Leading-edge flaps, called *slats*, may be added to increase lift even further. They are often fitted to combat aircraft, where the very small wing needed for good high-speed performance is highly unsuited to providing lift at low speeds for take-off and landing. When extended, many slats leave a gap or slot between the slat and the leading edge. This gap allows air to flow through and helps the airflow to stay attached to the wing surface, reducing the tendency for the wing to stall. Flaps and slats increase both lift and drag, and both are

Figure 6.12 Flaps and slats. *Flaps reduce the landing speed of the aircraft by increasing both lift and drag. Slats extend forward from the leading edge when deployed, and increase the camber or curvature of the wing section, which increases lift. Both flaps and slats can be deployed when needed, and retracted again for normal flight. The slat (in this case a Krueger flap) and double-slotted Fowler flap shown here are quite common on medium and large transport aircraft.*

an advantage for landings. When the aircraft speed is higher, they are not needed and are retracted out of the airflow.

Spoilers

Spoilers are fitted to the top surface of the wing of aircraft with very good glide and low-speed performance. During landing, aircraft of this type produce so much lift, even at low speeds, that they are difficult to land, because they tend to float. When operated, spoilers interfere with the normal air flow over the wing, increasing drag but reducing lift. This steepens the glide angle and makes accurate touchdowns easier.

Lift dumpers

Fitted to the top surface of the wings on larger aircraft (Figure 6.13), these devices are operated as the aircraft touches down. Deploying lift dumpers instantly reduces lift, in the same way as a spoiler, and prevents the aircraft bouncing back into the air. This makes the landing more positive, and puts the full weight of the aircraft onto the wheels very quickly. The aircraft wheel brakes are more effective, reducing the length of the landing run.

Figure 6.13 Lift dumpers. *Lift dumpers are usually operated automatically as the aircraft touches down. A large amount of the lift from the wings is quickly eliminated, which reduces the tendency of the aircraft to bounce, and increases the effectiveness of the wheel brakes.*

ICING

Because of the pressure changes (and hence temperature changes) which occur as the air flows over the wings, in certain conditions ice will form, particularly near the leading edge. Ice can add considerable weight to the aircraft, and changes the shape of the wing section from the carefully developed section seen in the wind tunnel, making it much less efficient. Ice build-up must be prevented or removed frequently, or handling problems will be experienced which may result in the loss of the aircraft. An anti-icing or de-icing system is fitted to almost all aircraft, and acts on wing leading edges and often on tail-surface leading edges and engine intakes. It is also often incorporated in propellers and in the carburettors of piston engines. De-icing and anti-icing systems are described in detail in Chapter 13.

CHAPTER 7

TAIL UNITS

Objectives: to describe the design features and functions of the parts of the tail unit, construction methods and the advantages and disadvantages of tail versus canard control layouts.

INTRODUCTION

If an aircraft is to be useful, it must be capable of flying from one point to another under complete control. This means that the pilot must be able to steer the aircraft to the required heading, and be able to control its attitude. The aircraft must be capable of withstanding gusts and turbulence to a reasonable degree. It must also normally be stable, so that it maintains the heading that the pilot sets, without needing constant movement of the controls.

In most aircraft, the sole function of the tail unit, or canard foreplanes and fin(s) for non-conventional layouts, is to provide this stability and control. Stability is the tendency of the aircraft to return to its original attitude following some disturbance, such as a gust. To allow it to be flown safely, without excessive demands on the pilot, an aircraft must not be unstable. In other words, it must not have a tendency to turn away from the course set by the pilot. In most cases, an aircraft must have positive stability – it must have a tendency to turn itself back towards straight and level flight following a disturbance. But in some aircraft, such as aerobatic sport aircraft or combat aircraft, too much stability is undesirable, and many are neutrally stable. This means that the aircraft is neither stable nor unstable, and continues along the path set by the pilot. Thus the pilot does not need to steer the aircraft against any tendency to right itself.

It is quite possible to fly a slightly unstable aircraft, but the pilot must constantly apply corrective control inputs as the aircraft tends to divert from the path set. For civil aircraft, this would be very tiring, unless the aircraft is fitted with an autopilot which is capable of maintaining the attitude required. If the aircraft is made less stable still, it becomes impossible to fly without some artificial means of restoring stability. So why should anyone wish to

75

design or fly an unstable aircraft at all? The answer is that unstable aircraft can turn extremely quickly, a definite advantage in air combat. The stability is generated artificially by using a computer control system to detect any diversions from the required path and to correct them. The speed of response of a computer means that a highly unstable aircraft can be flown in this way. When the pilot requires the aircraft to turn, the input to the control column is monitored by the computer and the appropriate commands are fed to the aircraft control surfaces. The result is an aircraft which appears to the pilot to be stable but highly responsive and manoeuvrable.

Since an aircraft flies in three-dimensional space, stability and control are required in three directions, each at right angles to the others. Any three such directions would be suitable, but it is most convenient to specify these axes in terms of *lateral* (left and right), *vertical* (up and down) and *longitudinal* (fore and aft). For aircraft turns, three corresponding manoeuvre cases are used:

- pitch – rotation of the aircraft about the lateral axis, raising or lowering the nose so that the aircraft will climb or descend
- yaw – rotation of the aircraft about the vertical axis, turning the nose left or right to change the heading of the aircraft
- roll – rotation of the aircraft about the longitudinal axis, raising one wing tip and dropping the other without changing the direction of flight

Pitch, yaw and roll manoeuvres (Figure 7.1) allow the aircraft to be turned in any direction, giving complete control over the attitude and direction of flight (provided, of course, that the aircraft is capable of flying in that attitude).

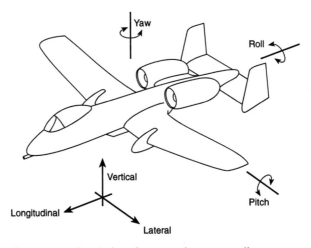

Figure 7.1 The three conventional aircraft axes and corresponding manoeuvre cases. *Any set of mutually perpendicular axes could be used, but for convenience we normally use the three shown here. Each manoeuvre case consists of rotation of the aircraft about one of the primary axes.*

Most aircraft manoeuvres use combinations of these basic cases, to provide smooth turns. Stability and control are normally provided by a separate set of surfaces for each of pitch, yaw and roll.

THE FIN AND RUDDER

The fin provides stability in yaw (left and right). If the aircraft heading is disturbed, by a sideways gust for instance, the fin will produce lift as it is turned away from the air stream. This lift will tend to turn the aircraft back towards its original heading. The effectiveness of the fin is given by the maximum size of this lift force multiplied by its moment arm – the distance from the aircraft centre of gravity. The fin is placed as far back as possible, so that the moment arm is maximised. This allows a relatively small fin to be used, reducing weight and drag.

When the aircraft is required to yaw, the rudder is deflected – to the left if the aircraft is required to yaw left or right to yaw right. This changes the normally symmetrical section so that lift is produced on the fin. As before, the lift turns the nose of the aircraft towards its new heading.

THE TAILPLANE AND ELEVATORS

The tailplane provides stability in pitch, i.e. up and down. If the aircraft pitch attitude is disturbed, the lift produced by the tailplane will change. This change of lift will tend to restore the aircraft to its original attitude. Like the fin, the tailplane is placed as far back as possible, so that the moment arm is maximised.

Figure 7.2 Conventional aircraft tail unit. *The conventional arrangement of the tail unit is used on almost all aircraft, and consists of separate surfaces to provide stability in pitch and yaw, each with a control surface.*

When the aircraft is required to climb or descend, the elevators are deflected. This changes the *camber* or curvature of the tailplane section, which changes the amount of lift it produces. If the lift is increased, the tail is lifted and the aircraft descends; if the lift is reduced, the tail drops and the aircraft will climb.

Many aircraft do not use separate elevators – the aircraft is controlled in pitch by moving the entire surface to vary the lift produced. The tail rotates about a pivot near its centre, and is usually moved by hydraulic jacks. This is known as an *all-flying tail*, which is stronger and more effective during high-speed flight. In combat aircraft an all-flying tail is less susceptible to battle damage.

CANARD FOREPLANES

As an alternative to a conventional tailplane and elevators described above, foreplanes may be placed at the front of the aircraft (Figure 7.3). They operate in a similar way but, because they are positioned near the nose of the aircraft, deflection of the control surfaces has the opposite effect. In other words, deflecting the control surfaces down (increasing lift) raises the nose and causes the aircraft to pitch up. Of course, the cockpit controls are still arranged so that pushing the control column forwards pitches the aircraft nose down, so the aircraft behaves in exactly the same way as far as the pilot is concerned.

Figure 7.3 Canard control surfaces. *An alternative, or sometimes a supplement, to the conventional tailplane position is to use canard foreplanes towards the nose of the aircraft. These can give improved turn performance and lower drag, but can suffer from problems with stall recovery. They are not common on civil aircraft, but when combined with fly-by-wire can offer significant performance advantages for fighter aircraft.*

An advantage of canard controls can be seen when the total lift produced by the wings and the control surfaces is considered. When conventional elevators are deflected upwards, the tailplane produces lift which acts in the opposite direction to the lift from the wing, so the *total* lift is initially reduced. As the tail drops, the angle of attack of the wing is increased and the lift produced by the wing is increased. The total lift will then be increased, causing the aircraft to climb. With canard controls, raising the nose is achieved by *increasing* the lift on the canard surfaces, making them more effective. For combat aircraft, this means that a tighter and more rapid turn is possible with canard control surfaces than with the conventional layout.

Another advantage comes about because of the requirement for natural stability on most aircraft. This requires that the incidence angle of the forward wing should be greater than that of the tailplane, and for a conventional arrangement, the centre of gravity has to be close to the centre of lift of the wing. Under cruising conditions, where the required angle of attack is relatively small, the effective centre of lift of the wing often moves aft of the centre of gravity. The tailplane thus has to produce a negative lift force, or down-force in order to keep the moments of the forces in balance, as shown in Figure 7.4. Since the aircraft must produce a total lift that is equal to its weight, the main wing thus has to produce extra lift to compensate for this down-force. All lifting surfaces produce lift-dependent drag, regardless of whether the lift is positive or negative, so the tailplane still contributes to drag, and the main wing produces additional drag as a consequence the extra lift. On a canard aircraft, both surfaces produce positive lift in the cruise condition, so they are helping rather than opposing each other, and the overall drag is less.

So why produce a conventional layout at all? The answer lies in the fact that in the canard configuration, the flow from the front wing tends to interfere with that on the main wing resulting in a lower ratio of lift to drag, and producing a number of problems due to buffeting and flow instability. On high performance military aircraft, these interference effects can sometimes be

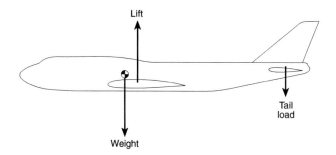

Figure 7.4 Lift and balance. *In cruise, the centre of lift of the wing often moves aft of the centre of gravity. To keep the moments of the forces in balance, the tailplane has to produce a negative lift force so that nose-up moment of this about the centre of gravity opposes the nose-down effect of the wing lift moment.*

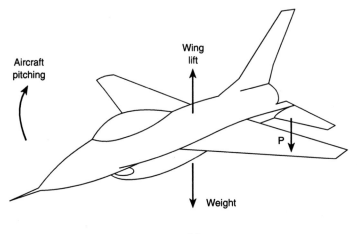

(a)

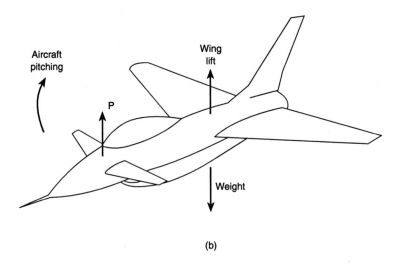

(b)

Figure 7.5 Control surface effect on lift. *To increase the lift on a wing, it must be given a greater angle of attack. On a conventional aircraft this is achieved by producing a negative lift force on the tailplane to pitch the aircraft nose-up, so there is an initial loss of lift. On a canard, the foreplane pitches the nose up by producing positive lift, so there is an immediate increase in lift.*

exploited to improve the manoeuvrability, but on civil aircraft, the problems are generally considered to outweigh the potential advantages. Very few canard arrangement civil aircraft have been built, and those that have, have not been a great commercial success. An exception is the range of light aircraft produced by Burt Rutan in the United States.

TRIMMING

If the position of the centre of gravity varies (for example as fuel is used up or different payloads are carried) or the aircraft speed is changed, the elevator position needed to maintain level flight will change. It would be undesirable for the pilot to need to apply a constant force to the controls, however small, to keep the aircraft flying straight and level. To allow for this, a small extra control surface is added to each main surface to allow the pilot to *trim* the aircraft. Trim tabs work independently of the main controls, although it is

Figure 7.6 Traditional fin construction. *Traditionally constructed aircraft have the fin built in a similar way to the wing, with several spars joined by ribs and covered with stressed skin.*

possible on many aircraft to fly the aircraft solely by moving the trim tabs, provided only gentle manoeuvres are required. Each trim tab is connected to a small wheel, the trim wheel, or a pair of trim switches for electrical trimming, in the cockpit. Trimming controls are provided for pitch, roll and yaw individually.

In many aircraft, including some airliners, conventional elevators are used to control the aircraft in pitch, but the pitch trim is adjusted by moving the entire tailplane, rather than by deflecting a trim tab. This is more complex than a trim tab, but reduces drag. A typical arrangement would consist of an electrical actuator attached to the front spar, and a hinge at the rear spar.

STRUCTURE OF THE TAIL

Traditionally, the structure of most tail units is often similar to that of a wing, but of course on a smaller scale (Figure 7.6). The tailplane and the fin contain a front and rear spar, connected by ribs to give the section its shape, and covered with a stressed skin supported by stringers. A separate leading edge is added, and the control surfaces are hinged from the rear spar, normally running the entire span of the unit. If the fin is highly swept, a multi-spar layout may be used.

In many modern aircraft, particularly of composite construction, new methods have evolved. A single spar may be used, with the tailplane or foreplane made entirely from a composite skin around a honeycomb structure. Less extreme variations are similar to the traditional structure but use composite materials, particularly for leading edges and tips where stresses are low.

CHAPTER 8

FUSELAGES

Objectives: to describe the function of the fuselage, the design features of typical aircraft fuselages and the components.

INTRODUCTION

The fuselage of any aircraft normally performs two main functions. Firstly it carries the payload, whether it is just the crew, in the case of a sport aircraft, or passengers or cargo for a transport aircraft. Secondly, it forms the main structural link in the complete assembly that is the aircraft. This structural link is responsible for carrying the wings and the tail unit in the correct aerodynamic positions, bearing the weight and other loads from all the components and transmitting the lift loads to support them. The fuselage often carries the engines and undercarriage. It is also responsible for providing a safe environment so that the crew and passengers can survive, and travel in comfort, in what would otherwise be very hostile conditions. So the fuselage must be able to resist high loads of various types. The weight and lift loads from the tailplane create large bending loads in the fuselage, whilst the fin and rudder create torsional loads that attempt to twist the fuselage. Cabin pressurisation creates a differential load due to the different internal and external pressures, which try to burst the fuselage. This pressure difference is typically up to 56 000 newtons per square metre (that is a force equal to the weight of six cars for every square metre of fuselage skin).

There are many other loads on the fuselage, some of which occur only during certain phases of operation. The structural strength and stiffness of the fuselage must be high enough to withstand these loads, so that every part of every flight may be carried out with a high degree of safety. At the same time the weight must be kept to a minimum, so that the operating efficiency of the aircraft is maximised. This means that any piece of structure that is not bearing a reasonable share of the load is adding unnecessary weight.

FUSELAGE STRUCTURE AND SHAPE

The fuselage is considered to be made in three sections:

- the nose section, or forebody
- the centre section or centre body
- the aft (rear) section or aftbody

The three sections will carry different loads depending on the role of the aircraft, but in all types the centre section needs to be very strong and stiff. In flight, the whole aircraft will be supported by lift from the wings, transmitted through the centre of the fuselage to carry all of the other parts. In transport aircraft, the majority of the fuselage is cylindrical or near-cylindrical, with tapered nose and tail sections. This is a convenient shape for carrying cargo or passengers, and makes it possible to *stretch* the aircraft (make the fuselage longer to increase its carrying capacity) by inserting extra pieces or *plugs* without a major redesign of the fuselage. Combat aircraft are quite different, and the shape of the fuselage can be quite complex, because of the special task it does.

There are other factors that must also be considered when the fuselage is designed. Doors must be added so that passengers or cargo can be loaded, and passenger aircraft will require windows. These increase the problems of weight – because they require cut-outs in the fuselage to allow them to be fitted they require extra structure to maintain the required strength, because the direct load paths are interrupted. It is extremely difficult to carry loads through windows and doors, so they are usually (but not always) regarded as non-load-bearing, and it is assumed that they contribute nothing to the required strength. The structure in the region of these cut-outs is therefore stronger and heavier than elsewhere. Most aircraft will require a level floor, to which seats or

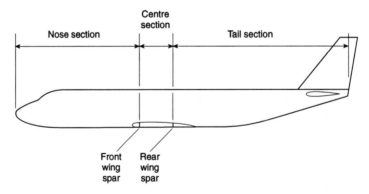

Figure 8.1 Three main sections of the fuselage. *The fuselage is made up of three main sections – the forebody, centre body and aftbody. The centre body carries the highest loads, and must be very stiff and strong to support these loads. The joint of the wing to the fuselage is the primary joint of the entire aircraft structure.*

load attachments can be fitted, and this too will impose loads on the structure.

Although the construction of a fuselage appears at first sight to be very complex, it consists of a small number of basic components, repeated at intervals until the required structure is complete.

As in wing construction described in Chapter Six, fuselage construction is usually of stressed-skin type, where the skin itself takes a proportion of the load. The semi-monocoque construction, which is virtually standard in all but the lightest of aircraft, consists of a stressed skin with added stringers to prevent buckling, attached to hoop-shaped *frames*. The designer will try to find the best compromise of skin strength (and weight) and frame and stringer strength, with the skin usually taking about half of the loads.

At the nose and tail, double curvature, like the surface of an egg, makes the skin even stiffer. A floor, comprising a number of beams across the inside of the fuselage and covered in sheet alloy or composite panels, leaves a flat surface for walking on and fitting seats. It also leaves a space below for luggage, freight and the many aircraft systems.

Pressure bulkheads (Figure 8.2) are fitted in the nose and close to the tail of most pressurised transport aircraft. They are flat discs, like a drum skin, or

Figure 8.2 Aft pressure bulkhead. *Pressure bulkheads define the foremost and aftmost regions of the fuselage that are pressurised above the outside air pressure. They may take the form of flat discs, as here, or a curved bowl. In either case, they will usually have some form of reinforcement to allow them to support the very high loads generated by the pressure differential across them.*

curved, like a breakfast bowl, and their job is to carry the loads imposed by pressurisation of the fuselage. As previously mentioned, cut-outs need to be made in all fuselages to allow for doors and windows, requiring local reinforcement. It is important to make sure that loads can be routed around these cut-outs, and spread evenly into surrounding skin and structure. The ideal shape for a cut-out in a fuselage is an ellipse, and many aircraft have windows this shape. This is not very practical for doors, though, and the more common arrangement is a rectangle with distinctly rounded corners.

FRAMES AND STRINGERS

The basic skeleton of a typical fuselage is made up of circular or near-circular frames, linked by stringers running fore and aft under the skin (Figures 8.3). The frames give the fuselage its cross-sectional shape and prevent it from buckling when it is subjected to bending loads. They protrude into the fuselage interior by between 50 and 150 mm, depending on aircraft size, leaving the rest of the fuselage clear for payload. Frames that need to be non-circular, such as those in the cockpit window area, must be much thicker and stronger than in

Figure 8.3 Frames and stringers. *Frames give commercial aircraft fuselages their circular or slightly oval section, the most efficient shape to resist the loads generated by pressurising the fuselage. Stringers give a large increase in the stiffness of the skin under torsion and bending loads, with minimal increase in weight. Frames and stringers make up the basic skeleton of the fuselage, although the extra strength added by the skin is considerable.*

other areas. Frames also assist in resisting pressure loads, which generate *hoop stresses*, stretching the skin. The stringers also support the skin under torsional and compression loads, preventing buckling. They do this by adding depth to the skin, with little increase in weight.

If the aircraft is not pressurised, a square or rectangular section is usually more convenient, since internal space can be used more effectively.

In combat aircraft, and also in unpressurised civil aircraft, the fuselage is not required to be of circular section, so the frames will be of a different shape. The pressurised region of a combat aircraft is restricted to the cockpit area, and pressure differentials are lower, so other requirements take precedence over these loads. It is accepted that a non-circular section here will require extra structure around this area, but the improved aerodynamics, weapon locations and other factors override the weight penalty incurred.

FLOORS

Because the cylindrical or near-cylindrical section most suitable for pressurised fuselages is not convenient for carrying passengers or cargo, a floor structure needs to be incorporated. This provides a level surface onto which the seats, cargo attachments and similar items may be attached. The floor must support the loads generated by these items, bearing not only the total weight but also high localised loads. The total loads appear primarily as bending moments across the floor. The local loads are concentrated onto small areas, which are mainly dissipated by the surface covering, often of a composite sheet such as Fibrelam.

The floor behaves as a beam running laterally across the fuselage, and is attached to the frames at each side. Longitudinal (fore-and-aft) members support and stabilise the lateral beams. As with any beam, it is important to provide sufficient depth in the structure to allow the bending loads to be carried without requiring a heavy structure. In some aircraft the floor is supported by additional struts or posts, which reduce the bending moments on the floor beams but may partially obstruct the area below the floor, and so reduce the usable cargo volume.

In aircraft with pressurised fuselages, the fuselage volume both above and below the floor is pressurised, so no pressurisation loads exist on the floor. If the fuselage is suddenly de-pressurised, for instance due to the loss of a lower door, the floor will then be loaded because of the pressure difference. The load will persist until the upper section of the cabin has equalised with the lower section, usually via the floor-level side wall vents. If the floor collapses, then a major catastrophe is almost inevitable, and at least one aircraft has been lost for this reason. So the designer is required to prove that the floor will not fail in these circumstances, and additional measures have been introduced to ensure that an aircraft cannot be flown unless the cargo doors are fully locked.

Seats are mounted on the floor using *seat tracks*, which are made in standard dimensions. Galleys and toilets are also designed to be fitted to the seat tracks.

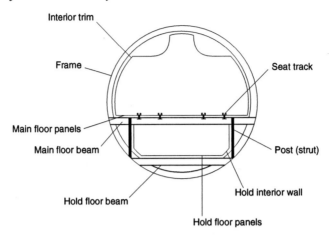

Figure 8.4 Airliner floor design. *The floor of passenger and cargo aircraft consists of lateral beams attached to the frames, and in some cases given additional support by posts or struts. The cargo area in the hold will also have interior walls and a floor to ensure that cargo does not rest directly on the skin.*

These features increase the flexibility of layout, so the operator can easily tailor the layout to suit the requirements of a particular route or type of operation.

If an aircraft is to be used as a pure freighter, or as part freighter and part passenger aircraft (the *combi* layout), the area by the freight door is often fitted with a *ball mat*, and rollers are set into the floor along the rest of the cargo area (Figure 8.5). The ball mat is a set of ball bearings set into the floor that allow easy motion of cargo in any direction. The cargo must of course be properly restrained for flight, and there is usually some form of standardised load-restraint system that will be chosen to give maximum flexibility of use. Most air cargo is carried on pallets, and many cargo aircraft have latches fitted to the floor that can clamp the pallets very quickly and easily.

Surprisingly, most air freight is not carried in specialised freight aircraft, but on scheduled passenger services, utilising otherwise unused space in the hold below the main floor. Again, pallets are used for most freight, and the hold will normally have a flat cargo floor fitted for this purpose.

DOORS AND WINDOWS

An aircraft would be of little use if it had no provision for loading and unloading passengers, crew or cargo, so doors are obviously required. Similarly, windows of various types will normally be needed. Doors and windows present particular problems to the designer, since they are usually unsuitable to carry many of the loads that are present on the remaining structure. Doors must be capable of withstanding the pressure loads, of course,

Figure 8.5 TriStar cargo variant floor details. *The floor of cargo aircraft is often fitted with a ball mat and a set of load restraint latches. These allow easy and fast loading, and securing of palletised and unpalletised loads with maximum flexibility.* Photograph courtesy Marshall Aerospace Ltd.

and must not be permitted to open in flight. However, they must be easily opened in an emergency on the ground, and passenger aircraft must have at least a minimum specified number and type of doors to allow rapid evacuation when required.

A typical freighter aircraft will have a much larger door than a passenger aircraft, and this may be either a side- or top-opening door, depending on the aircraft. Large aircraft usually have a door that opens upwards, using hydraulic or electrical power. It can form a canopy over the loading area if opened to the horizontal position, or can be fully opened for maximum clearance. Because of the size of these doors, it is necessary for them to transmit some of the loads from the frames and stringers, to maintain continuity. Where doors are smaller, the surrounding structure is reinforced to transmit the loads around the door. Passenger doors on all pressurised aircraft must be of the plug type, where the door arrangement is such that internal pressurisation holds the door firmly closed in flight (Figure 8.6). The door may be opened by sliding upwards inside the fuselage or by lifting upwards before opening, but the most common arrangement is a door that initially is pulled inside the fuselage. The door is then turned and passed through the door aperture, so that in its fully open position it is outside. Gates at the top and bottom of the door fold to reduce the door height and allow it to pass through the aperture. This complicated arrangement allows the most unhindered passenger exit, whilst still allowing it to operate as a plug door. An inflatable chute operates automatically if the doors are opened in an emergency, so that passengers can reach the ground quickly and safely.

Cockpit windows must not only provide adequate visibility to the crew, but must also be capable of withstanding bird strikes. Even a small bird impacting

Figure 8.6 Passenger door. *Doors on all modern passenger aircraft are of the plug type, where the pressurisation holds the door firmly closed even if the latches fail. This type, on the Boeing 757, is the most common – the door is pulled into the fuselage, rotated and then passes through the aperture to reach the open position. The pressure loads are transmitted into the fuselage through the hinges and latches.*

at the speeds at which modern aircraft fly contains a large amount of kinetic energy. Penetration must be prevented, and visibility maintained as far as possible to allow safe return or continuation of the flight. The windows are exposed to external air temperatures that, particularly in high-speed aircraft, can vary within a very large range, and must retain the high strength required to contain pressurisation under these conditions. Glass used in most windows has a good performance over a wide temperature range. The stress applied to the plastic canopies used in many military aircraft needs to be kept low, because they are often prone to degradation over time. In most aircraft, the windows are of a fail-safe construction, with at least two panes, and will maintain adequate strength even if one pane fails. Passenger cabin windows have an additional, non-load-bearing pane so that the structural layers cannot be weakened by scratching from inside the cabin. Cockpit windscreens and fighter canopies are normally laminated from a number of layers, often combining different materials to achieve the required overall properties.

To prevent condensation, cockpit windows on most aircraft are heated, often by incorporating a gold film through which an electric current is passed. This film is only a few atoms thick, so it is transparent, with an amber tint the only visible evidence of its existence.

Since doors and windows need cut-outs in the fuselage wall to accommodate them, arrangements have to be made to support them on the fuselage structure,

Figure 8.7 Jetstream window detail. *The cut-outs to accommodate the windows mean that the structure around them needs to be reinforced to maintain adequate strength. Note the beams above and below the window, and the fact that the window is located between the fuselage frames to maintain continuity of the structure as far as possible.*

and to compensate for the loss of strength caused by the aperture. Door apertures have a sill beam at floor level, and a similar beam over the top of the door. These beams maintain the shape of both the door aperture and the pressure shell of the fuselage. The fuselage frames on either side of the door are reinforced also, so that loads can be carried around the door, and also to carry the loads from the door hinges and latches.

Windows also need reinforcement around them, but not to the extent of that for the door. It is usual to place windows between fuselage frames, so the disruption is limited to the skin and perhaps one or two stringers. Localised reinforcement takes account of the pressure loads transmitted from the windows themselves.

FATIGUE PROBLEMS

The stresses imposed on a fuselage in flight and on the ground are cyclic, which is to say that they vary repetitively. Tensile loads from pressurisation loads and bending are additive along the top of the fuselage in flight, and bending loads from the wing are transmitted to the fuselage frames near the wing attachment. Both of these areas are critical in fatigue (see Chapter Five), and the choice of materials and structure in these areas is extremely important.

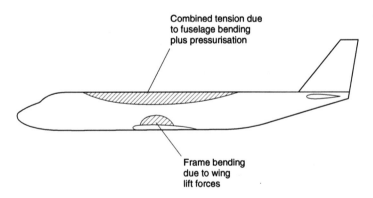

Figure 8.8 Fatigue-critical areas of the fuselage. *Tensile loads as a result of fuselage bending and pressurisation loads combine in the top of the fuselage to give high cyclic stresses, and therefore a fatigue hazard. High bending loads from the wing root combine with pressurisation loads to create a similar critical area at the joints of the fuselage frames to the wing spars.*

CHAPTER 9

ENGINES

Objectives: to describe the types of engine and the locations on an airframe where engines can be installed, list some of the advantages and disadvantages, and detail various engine-mounting arrangements.

INTRODUCTION

A wide range of engine types are used in aircraft. Each type has particular advantages and drawbacks, and its performance in terms of power output or thrust and fuel consumption is critically important in most cases. For a particular type of aircraft, there is often a single engine type that is most suitable, so the choice is primarily concerned with specifying the manufacturer and model.

Also of importance are the location and installation of the engine or engines on the aircraft. The choice of installation is influenced by the type of aircraft, the type of engine and the way in which the aircraft will be used in service.

TYPES OF ENGINE

Aircraft power plants fall into five main types:

- turbo-jet
- turbo-fan
- turbo-prop
- prop-fan
- piston engine

The first four on the list are variants of the gas-turbine engine, which is capable of delivering high power in an extremely compact size and with low weight. Each variant is most suited to a particular aircraft flight speed. The operating efficiency (loosely defined as power absorbed divided by the rate of fuel burn) is maximised when the velocity of the air expelled from the jet, fan or propeller is close to the speed of the aircraft. This means that an aircraft such as an airliner that flies at sub-sonic speeds cannot be efficiently powered by an engine such as a turbo-jet, which has a high exhaust velocity.

Turbo-jet engines produce all of their thrust as a high-speed gas stream of relatively small diameter, because all of the exhaust gases have passed through the combustion process. This is ideally suited to very high-speed aircraft – it may be the only viable power plant.

Turbo-fan engines are based around the pure turbo-jet, but differ in that some of the exhaust gases are made up of air that has by-passed the engine core or gas generator (hence the term *by-pass engine*) and passed only through a fan. So the thrust is produced by an air stream of larger diameter than the turbo-jet, but at a lower velocity. The turbo-fan engine is suited to aircraft that fly in a range of speeds from around Mach 2 down to around 350 km/hour, depending on the by-pass ratio. The by-pass ratio is the ratio of the air that passes only through the fan to that passing through the core of the engine, providing the oxygen required to burn the fuel. The higher the by-pass ratio (i.e. the higher the proportion of air passing only through the fan), the lower the aircraft speed at which the engine will be efficient, but the larger the engine's diameter for a given thrust.

At lower speeds still, even the turbo-fan becomes less efficient, and the propeller comes into its own. The **turbo-prop** engine still takes advantage of the high power capability and low weight of the gas turbine engine, but uses the power produced to turn a propeller via a gearbox. The small diameter of the engine itself in comparison with a piston engine of the same power output is a distinct advantage, allowing streamlined engine installations.

The **prop-fan** engine is a very new development, and seeks to gain the advantages of the turbo-fan and turbo-prop engines. The fan duct, which makes up the outer wall of the turbo-fan, is omitted in this design, and the exposed fan combines features of both a fan and a multi-blade propeller. The fan may run at high speed, driven directly from the turbine, or may run more slowly, via a gearbox. The fan is normally two-stage (two sets of blades one behind the other), and contra-rotating. These engines are still under development, and technical problems are progressively being overcome.

For the smallest aircraft, a gas turbine engine is too expensive, both to buy and to maintain, so **piston engines** are the most common choice. They are cheap, simple and reliable, require little in the way of specialist servicing and provide adequate power for most light aircraft.

ENGINE LOCATIONS

Once the type of engine has been selected, the location of the engine or engines in the airframe is of prime concern. If only one piston engine or turboprop engine is fitted it will normally be in the nose. If a single turbo-jet or turbo-fan engine is used it will also normally be in the fuselage, but further aft, near the centre of gravity. If there are two or more engines there is a variety of positions to be considered.

With turbo-prop and piston engines, the diameter of the propeller places severe limitations on their location; twin- and four-engined turbo-prop aircraft will almost inevitably require the engines to be wing-mounted. Turbo-fan, turbo-jet and prop-fan engines allow greater flexibility in their location (Figure 9.1).

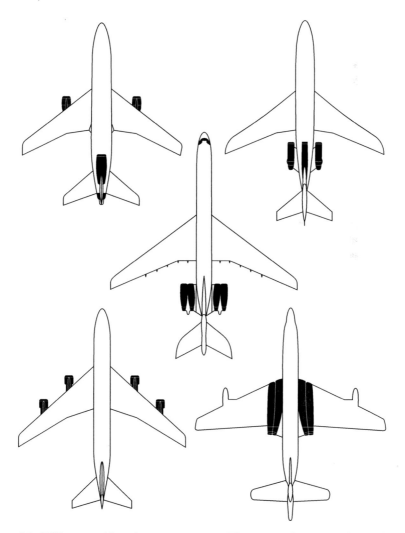

Figure 9.1 Different multi-engine arrangements. *There is a wide variety of possible engine arrangements that can be used in aircraft. This figure shows common arrangements for triple- and four-engine aircraft. All have particular advantages in certain applications, and the designer chooses the arrangement that is most suited to the specific aircraft type and usage.*

Most engines have similar mounting points provided regardless of where they are to be located (Figure 9.2). The most common arrangement is for the major thrust and weight loads to be borne at two primary locations at the top of the engine, with stabilising links to prevent rolling, or at both sides and the rear. The way in which these points are connected to the aircraft *structure*, however, varies widely depending on the type of installation.

ENGINES INSIDE THE FUSELAGE

Combat aircraft use mainly external installations for the weapon systems, and the fuselage is not required to carry an internal payload, so it is an ideal location for the engines. This gives the aircraft smooth exterior lines, for low drag, and lends the engines some protection from battle damage. The thrust must be large in relation to the aircraft weight, so the engines are relatively

Figure 9.2 Engine mounting points. *Engines of similar type and size tend to adopt similar mounting arrangements, simplifying the task of incorporating a variety of engines in a given aircraft. This gives purchasers a choice of engine, and maximises the sales potential of both aircraft and engine.* Photograph courtesy of Rolls-Royce plc.

large, and take up the major portion of the fuselage volume. Intakes (Figure 9.3) must be designed to supply large volumes of air to the engines, and the shape of these can be extremely critical to provide the best possible performance over a wide range of speed, angle of attack and altitude. In twin-engine installations, the engines are close together, reducing the adverse yaw effects if an engine fails in flight.

One of the most remarkable aircraft with an internal engine is the British Aerospace Harrier. Using vectoring nozzles for both the by-pass air and the engine exhaust, coupled with an engine that can produce more thrust than the weight of the aircraft, the Harrier can land and take off vertically when lightly loaded. Even with a heavy weapon load, the take-off run is substantially reduced, and vectoring the nozzles in forward flight (called *VIFFing*) allows a superior turn performance over most normal fighters. The secret of the Harrier's success is the Rolls Royce Pegasus engine.

Most light aircraft, and some small turbo-prop aircraft, have a single engine mounted in the nose (Figure 9.4). The engine is supported by a welded framework cantilevered from a firewall, or bulkhead, immediately forward of the cabin. The firewall, as its name suggests, separates the hot engine from the rest of the structure. The engine mounting attaches directly to the flat firewall, so a different engine can often be fitted simply by replacing the engine-

Figure 9.3 Engine intakes. *The air intakes are critical to obtain good performance of the engines under all possible flight speeds, angles of attack and altitudes. Note that the intake is set slightly away from the fuselage, so the fuselage boundary layer is diverted away from the intake. This gives a smoother air flow into the engine.* Photograph: Alistair Copeland.

Figure 9.4 Small turbo-prop engine-mounting arrangement. *This engine is supported by a welded tubular framework, cantilevered from the firewall. The horseshoe-shaped member curving over the engine at the forward end attaches to the forward mounting points, and is diagonally braced for lateral stiffness.*

mounting framework, provided the weight is not substantially different. This allows a manufacturer to update the design to incorporate a more powerful or more economical engine, perhaps giving an ageing design a new lease of life.

EXTERNALLY MOUNTED ENGINES

Most transport aircraft have externally mounted engines, leaving the fuselage interior volume clear for carrying passengers or freight. The designer has a choice of locations for turbo-fans and prop-fans, as already stated, but turbo-props are essentially limited to wing locations. Engines may be mounted on the rear fuselage, or they may be mounted on the wings, generally in under-slung pods. Each location has particular advantages and disadvantages, and the designer has to balance these when selecting the location.

Rear-mounted engines – advantages

- less thrust asymmetry if an engine fails
- engine is well clear of ground, reducing chance of debris and runway water ingestion

- penetration of pressure shell unlikely if turbine disc breaks up (provided engine is behind rear pressure bulkhead)
- close grouping of engines eases cross-flow of air for starting
- convenient air-supply routing for cabin conditioning air

Disadvantages

- reduced accessibility for servicing
- turbulent air flow into engines due to presence of fuselage
- often higher cabin noise levels

Wing-mounted engines – advantages

- weight and thrust of engines reduces bending and torsional loading on wings in flight, so a lighter structure may result
- easy and rapid access for servicing, particularly at the departure gate
- clean air flow into engines
- easy to modify the design to use an alternative engine
- convenient routing for fuel lines

Disadvantages

- close proximity to ground increases risk of damage due to ingestion of debris and from ground impact during landings, and water ingestion from wet runways
- shear-off system is normally required for emergency landings, with increased risk of premature failure
- increased risk to ground personnel during aircraft turn-round
- high thrust asymmetry if an engine fails, requiring large rudder deflections to trim
- taller and more complex undercarriage required to give adequate ground clearance

Twin- or multi-engined propeller-driven aircraft *must* have their engines spaced out along the wing to provide clearance between the propeller tips and the fuselage. The closer the tips are to the fuselage, the more noise is generated inside the fuselage, and the further away they are, the more the aircraft yaws if an engine fails. The radius of the propeller also creates ground-clearance difficulties with low-wing arrangements in some cases, and high-wing turbo-props (Figure 9.7) are common in the medium-size category. The high-wing arrangement allows the aircraft to sit close to the ground, which makes loading easier, especially when the aircraft has a rear loading door. It can cause problems with undercarriage design, though, and the choice is usually between long, spindly main units that retract into the engine nacelles and short, stocky units that retract into the fuselage. To reduce the effects of intrusion into the

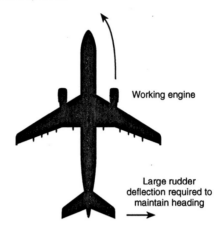

Working engine

Large rudder
deflection required to
maintain heading

Figure 9.5 Effect of asymmetric thrust due to engine failure. *If a wing-mounted engine fails, the thrust from the remaining engine or engines causes the aircraft to yaw strongly. This needs a large rudder, capable of large deflections, to allow the aircraft to be kept in straight flight.*

load space, the fuselage may have bulges or *panniers* into which the undercarriage can be retracted, which also give the undercarriage a wider track, improving stability on the ground.

Figure 9.6 Engine on wing-mounted pods. *Mounting engines in pods on the wings is almost universal for medium and large passenger and freight aircraft. Access for servicing is simple, and it allows an aircraft manufacturer to offer an aircraft with a choice of engines, to suit most operators. Note how close the engine is to the ground, making the engine very prone to damage from debris during movement on the ground.*

Figure 9.7 High-wing turbo-prop freight aircraft. *The arrangement of the Hercules shown here is typical of many turbo-prop aircraft in this category. The high-wing arrangement allows generous propeller ground clearance, permitting a short undercarriage to be used. This makes loading easier, via the rear loading door.* Photograph: Alistair Copeland

FUEL SUPPLY

Wherever the engines are located, they must be supplied with fuel. To do this, a system of fuel pumps is incorporated to carry fuel from feeder tanks to the engines. In twin- or multi-engine installations it is a requirement that the fuel supply can be maintained if any component fails. This is achieved by having two or more fuel pumps, each supplying two or more engines, so that each engine is not dependent on a single pump. The maximum degree of safety and flexibility is provided by having the capability of drawing fuel from more than one tank. The fuel pumps are electrically driven, and again this supply would be derived from more than one source, so that even failure of the local electrical supply would not affect the entire fuel system.

VARIABLE-PITCH AND FEATHERING OF PROPELLERS

To provide good performance and economical operation over a range of aircraft speeds, propellers fitted to many aircraft have the capability of varying their pitch (the angle at which the blade meets the oncoming air) in flight. Varying the pitch is very much like changing gear in a car. Fine pitch is similar to low gear, giving good acceleration at low forward speed, and is used during take-off and climb. Coarse pitch is similar to top gear, giving economical operation at high speeds. If the propeller pitch is controlled automatically, an engine can be operated at constant speed. This is ideal for turbo-prop engines, since they operate most efficiently in a narrow speed range. Moving the throttles on such a system increases the power setting of the engines, but the engine speed does not increase because the propeller pitch is automatically coarsened to absorb the extra power. In simpler arrangements, the propeller pitch is controlled manually, either over a continuous range or perhaps between two settings, fine and coarse.

If an engine fails, the propeller will *windmill*, taking energy from the air stream and turning the engine. This causes extra drag, and may further damage the engine, so the propeller is *feathered* – the blade pitch is changed until the blades sit approximately in line with the air stream, and windmilling is prevented.

Figure 9.8 Thrust reversers deployed. *Clearly shown here are the thrust reversers in the deployed position. The two large 'buckets' rotate from their parked position to deflect the engine thrust forwards. This form of braking is effective even before the weight of the aircraft is fully on the wheels, and is a useful supplement to wheel braking, especially on wet or icy runways. The flaps on this aircraft have been deflected to the braking position (70°) as well – this reduces the lift produced, and greatly increases the drag.* Photograph courtesy Raytheon Corporate Jets Inc.

REVERSE THRUST

Aircraft wheel brakes are effective only when the full weight of the aircraft is on the main undercarriage. On wet or icy runways their effectiveness is reduced even then, so it would be useful if the thrust of the engines could be used to slow the aircraft after touchdown. Thrust reversers divert the thrust of gas-turbine engines forward, and provide braking which can be effective almost the moment the aircraft touches down. The same effect can be achieved with propellers by reversing the pitch of the blades. Reverse thrust supplements the wheel brakes, and allows the aircraft to slow to taxying speed more quickly, reducing turn-round time. It also allows a shorter runway to be used, which means that an aircraft can use airfields that would otherwise be unsuitable, and reduces brake wear and hence maintenance costs and down-time. So for civil operators, operating costs can be cut and access is opened up for additional services, and for military operators the operating effectiveness and flexibility of the aircraft are improved.

For safety reasons, it is important that the thrust reversers are inhibited in flight, which has been the cause of accidents in the past. An inhibitor system prevents the engagement of reverse thrust until the aircraft is safely on the ground.

CHAPTER 10

UNDERCARRIAGES

Objectives: to describe the job an undercarriage does, the importance of the undercarriage layout, the method of absorbing shock, the different types of main-wheel unit, retraction of undercarriages, and methods of braking and tyres.

INTRODUCTION

With very few exceptions, all aircraft need an undercarriage. This performs two main functions:

- It supports the aircraft on the ground.
- It absorbs the shock of landings and provides smooth taxying.

There is more to an undercarriage than just carrying out these functions, however. It must support the aircraft in the desired attitude on the ground, so that the drag on the take-off run is minimised, and the aircraft taxies without any tendency to float at normal speeds. It must withstand the loads that will occur during all movements on the ground, including braking and side loads. The undercarriage serves no function at all during flight, so it must be as small and light as possible.

UNDERCARRIAGE LAYOUT

There are many different layouts of undercarriage in current use. The type chosen depends on the type of aircraft and its intended use. For almost all aircraft, except some light aircraft, the tricycle layout (Figure 10.1) is preferred, because it supports the aircraft in a horizontal attitude, giving low drag during the ground run. However, there are several different kinds of main unit, for different installations.

The designer's main concern when choosing the type of main unit is how many wheels the unit will have, and their arrangement. This will depend on the weight of the aircraft and the way in which the undercarriage is to be retracted.

Figure 10.1 Tricycle undercarriage layout. *The tricycle layout is the most common arrangement for aircraft of all sizes. The main landing loads are taken by the main units, which contain large shock absorbers to absorb and dissipate the high kinetic energy due to the vertical motion of the aircraft. On touchdown, the aircraft pitches forwards until the nose unit touches down, and then the aircraft can be braked quickly to taxying speed under full control.*

Figure 10.2 Dual unit. *This arrangement is common on main units of combat and medium-sized commercial aircraft, and on nose-wheel units of larger aircraft. Its advantages lie in its compact dimensions, coupled with reduced wheel loading compared with single-wheel units, and its symmetry about the undercarriage leg. Nose units usually retract forwards, so that extension in an emergency is assisted by drag. Main units may retract in a variety of ways depending on the installation.* Photograph: Alistair Copeland.

Each main-wheel unit may contain a single wheel, a pair of wheels side by side or in tandem, or four or more wheels. As aircraft become heavier, the loading on each wheel increases, leading to a considerable increase in the damage done to runways. By having the weight spread over a greater number of wheels, the contact pressure of the undercarriage is reduced. This also increases safety if a tyre bursts on landing. The Boeing 747 has 18 wheels – four main units, each with four wheels, and a dual nose-wheel unit.

Apart from the single-wheel main unit, the simplest type is the twin-wheel side-by-side (or dual) arrangement, which is used on many fighters, as well as medium-sized transports such as the Boeing 727 and 737, the Fokker F28 and many turboprop aircraft.

By far the most common arrangement of main units for large aircraft is the dual-tandem layout, also known as a *bogey* or *truck* (Figure 10.4). This is widely used on commercial aircraft, since it gives a good combination of low ground pressure and relatively easy retraction arrangements. The Boeing 747, 757, 767 and the Airbus series are just a few examples of the many aircraft using this arrangement. It is easily capable of retracting forwards or sideways, and the bogey can be rotated to fit into awkward spaces. If necessary, the bogey can be held parallel to the ground during retraction, to allow a shallow well to be used.

Figure 10.3 Boeing E-3 undercarriage arrangement. *The undercarriage layout shown here is typical of many medium-sized transport aircraft. The main units are four-wheel bogeys, and a twin nose-wheel is used.*

Figure 10.4 Dual tandem (bogey) main unit. *The four-wheel bogey unit (also sometimes called a truck) is found on many aircraft of medium and large size, offering a reasonable wheel loading coupled with a wide range of options for retraction geometry and good access for servicing. The usual method of retraction is to retract the leg sidways into a well in the wing and fuselage.*

One of the main reasons for the particular choice of undercarriage arrangement is the problem of retraction. The main units of low-wing aircraft are usually retracted into the wing, which is quite straightforward in most cases. With high-wing aircraft, this would require a long undercarriage, which increases weight. Twin turboprop aircraft have engine nacelles on the wing, and it is quite common to retract the main legs into these nacelles. Otherwise, they must be stowed in the fuselage. However, the points of contact of the undercarriage with the ground must be far enough apart to make the aircraft stable during take-off, landing and taxying, so the shape of the main units can become quite complex.

The tandem undercarriage (Figure 10.5) is rarely used. However, a variation of the tandem arrangement is the jockey unit (Figure 10.6), which comprises two or three levered legs in tandem on each side of the fuselage, sometimes sharing a common horizontal shock absorber. It is particularly useful for high-wing medium-sized transport aircraft, because the undercarriage is easily retracted into *panniers* – bulges on the side of the aircraft. This gives a constant width of cargo area in the fuselage, and of course the widest load that can be carried is often restricted by the narrowest point in the load space. Among the advantages of this design are excellent rough-field performance and the ability to 'kneel' the aircraft by partially retracting the undercarriage to reduce the slope of loading doors. This is particularly useful where the aircraft is used to transport vehicles. The units also retract into a small space, without

Figure 10.5 Tandem main unit. *The tandem main unit is not widely used, except in some jockey units, but can be an advantage when there is a need to combine relatively high undercarriage loads with a very shallow undercarriage well.*

penetrating into the load space. This makes this arrangement ideal for transport aircraft like the Hercules.

Figure 10.6 Jockey unit. *The jockey unit is similar in action to the tandem main unit, but the retraction geometry is such that it can be retracted forwards or rearwards into the fuselage with minimal intrusion into the load space. The long, thin shape of the units after retraction is ideally suited to fit into bulges or panniers on the fuselage, so that the full width of the load space is maintained. Some aircraft allow the undercarriage to be partially retracted on the ground, as here, to reduce the slope of a rear loading ramp.*

There are a number of other wheel arrangements in use, including tri-twin tandem, dual twin, dual-twin tandem and twin tricycle, but the more complex the type the less commonly it is used. However, as increasingly large aircraft are developed to take maximum advantage of crowded airspace, the number of wheels in undercarriages must be increased to keep ground pressures reasonably low, and limit damage to runways and taxiways.

With combat aircraft, the main undercarriage has another limitation, which is the requirement to clear stores fitted under the fuselage. The undercarriage must not interfere with these stores either in its extended position or during retraction. Many combat aircraft carry under-fuselage stores, and this can result in some rather awkward-looking undercarriage arrangements.

The undercarriage design will normally allow for steering, and a reasonable turn radius is needed for ground manoeuvring. At the same time it must have a safety mechanism that prevents the nose wheel from being turned after retraction, and ensures that the wheel is straight for landing.

If the undercarriage hits a large obstacle that the wheels cannot climb, there is a risk that considerable damage may be done to the structure that supports the undercarriage. *Shear pins* are fitted, which will fail and allow the collapse of the

Figure 10.7 Combat aircraft undercarriage layout. *Undercarriage layout is constrained by the need to retract it into a very small, and often inconveniently shaped, space. This is further complicated by the heavy loads and high landing speeds of modern aircraft. Stability on the ground is yet another requirement. A military operator will usually also require the aircraft to be capable of taking off from repaired or adapted surfaces. All of these conflicting requirements complicate the designer's task, and can lead to unusual configurations.* Photograph: Alistair Copeland.

undercarriage before the load rises beyond a safe level. The aircraft will still be damaged, of course, but not to the same extent as it would without this feature.

The position of the undercarriage units is very important, particularly the main units (Figure 10.8). If they are too far forward, the aircraft may tip during loading and taxying. If they are too far aft, the aircraft will pitch forward violently during landing, which could cause the nose leg to collapse. If the main units are not sufficiently wide apart, the aircraft may tend to roll sideways on the ground, especially in side winds and during taxying. If they are too far apart, the aircraft may be prone to ground loops – a sudden violent turn to left or right, perhaps even more than a full circle. The nose leg must also be positioned carefully because its distance from the main units affects the proportion of the total weight that it carries. If it is too lightly loaded, the steering may not be effective, but the load must not be so high as to require the nose leg and associated structure to be unnecessarily strong and heavy. The designer will often be limited by the available structure and, as always, the position may be a compromise.

SHOCK ABSORBERS

When an aircraft lands, a large force is generated on the undercarriage as it meets the ground. This may be up to three times the weight of the aircraft for transport aircraft, and up to eight times for an aircraft landing on a carrier deck. To prevent damage to the structure, this shock must be absorbed and dissipated by the undercarriage. It absorbs energy using some form of spring, and dissipates it as heat. The energy must be dissipated, or it will be returned to the aircraft by bouncing it back into the air.

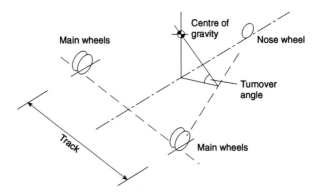

Figure 10.8 Undercarriage stability. *The fore-and-aft position of the main units affects the tendency to tip, and so limits allowable centre of gravity positions. It also influences the loads that are applied to the nose wheels during landing. The lateral position of the main units (track) affects the turnover angle, and hence dictates the tendency of the aircraft to roll during taxying. A low CG position improves stability.*

Figure 10.9 Light aircraft undercarriage. *Non-retracting undercarriages fitted to light aircraft are usually very simple, typically with the spring provided by bending of the leg, and damping, if any, by a rubber block.*

Several methods are used to absorb taxying and landing shocks. On light aircraft, the undercarriage may be just a piece of spring steel, with perhaps a rubber mounting in the aircraft fuselage. Rubber can be used both to provide the spring and dissipate the energy through *hysteresis*, in which the rubber converts the kinetic energy of the aircraft into heat energy.

With larger aircraft, rubber shock absorbers are not suitable, and a combined compression strut and damper assembly are used. The spring is provided by compressing either gas (normally nitrogen) or oil. Oil is used in both types to dissipate the energy and prevent rebound. The gas and oil type is called an *oleo-pneumatic* shock absorber (Figure 10.10), and the oil-only type is known as *the liquid-spring* (Figure 10.11). Except for the spring medium, both types operate in a similar way, and are efficient, compact and reliable.

It takes a great deal of pressure to compress oil by any significant amount, so operating pressures for the liquid spring are high and volume changes are quite small, providing a compact design. As the leg is compressed, oil flows through a relief valve, and through a leak hole in the piston head into the upper chamber. On rebound, the oil flows back through the leak hole and the recoil valve, which is more restrictive than the relief valve, producing a more gradual extension. In flowing through the valve, the oil absorbs energy, rising in temperature as it does so. By varying the flow rates of the valves, the compression and rebound characteristics can be tuned to the particular requirements; because the relief valve is spring loaded, it can be set to open only at relatively high loads. This has the effect of increasing the stiffness of the leg at low loads and reducing stiffness at high loads. This is particularly useful for nose legs, where the high stiffness prevents excessive pitching during taxying, but at the same time a soft shock absorber is provided to absorb landing shocks. In all types of undercarriage, deflection of the tyres absorbs some minor shocks.

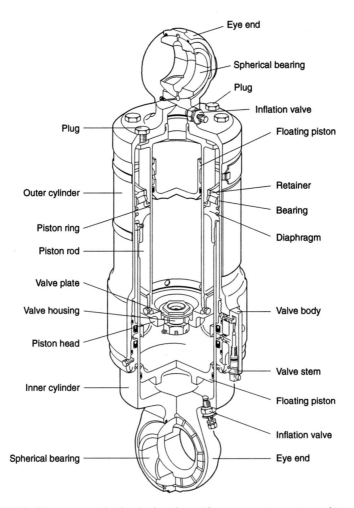

Figure 10.10 Oleo-pneumatic shock absorber. *Oleo-pneumatic struts combine the spring and damper in a single unit. The spring action is achieved by compressing a gas, usually nitrogen. Motion of the piston in the cylinder causes oil to pass between the top and bottom of the piston, and motion in each direction is controlled by a valve. The shock absorber is allowed to compress relatively easily, but the rebound rate is more limited, dissipating energy by heating of the oil. Thus the energy is not returned to the aircraft, which would otherwise cause it to bounce.* Courtesy Messier-Dowty.

The circular nature of the sliding piston and cylinder means that the lower section of undercarriage would be free to rotate about the axis of the leg, so it must be restrained. This is achieved by adding a torque link (Figure 10.12), which allows the leg to compress freely but stops the piston turning within the cylinder. This keeps the wheel pointing in the direction of travel.

The operation of nose-wheel undercarriage units is similar to that of main units, but their construction differs slightly in that they are usually designed to

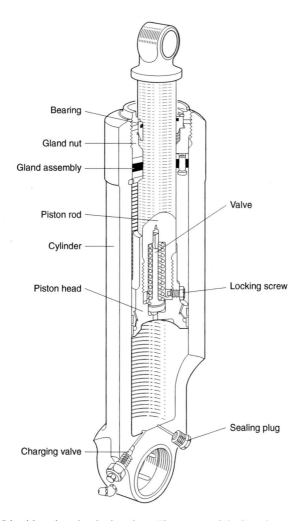

Figure 10.11 Liquid-spring shock absorber. *The action of the liquid spring unit is very similar to the oleo-pneumatic unit, but in this case the oil acts as both the spring and the damping medium. Because oil is much less compressible than gases, the unit can often be more compact, although this is partially offset by higher structural weight in the unit due to higher operating pressures.* Courtesy Messier-Dowty.

Figure 10.12 Torque link. *The torque link prevents rotation of the lower section of the undercarriage leg, which would allow the wheel to turn away from the direction of travel, whilst still allowing the oleo to extend and compress.*

allow the nose wheel to be steered. Nose-wheel steering can be achieved either by rotating the entire unit or by the use of steering motors on larger aircraft. A steering actuator is placed below the torque link, in order that the nose wheel can be turned when required to permit manoeuvring on the ground. For ground towing, it is often possible to disconnect the torque link or steering motors, so that the wheel will castor.

RETRACTION

An undercarriage would cause a lot of drag in flight, so it is retracted into the wings or fuselage of most aircraft when not required. In most cases, a hydraulic jack is used to pull up the undercarriage legs, about a pivot at the top.

Retracting an undercarriage can present a difficult design problem, since space must be found not only for the undercarriage leg and wheels, but also for the retraction mechanism. Careful design of the retraction arrangement minimises the size and weight of the jack. Strong points must be correctly located on the aircraft structure for the undercarriage and jacks. The leg must also be properly braced, to prevent collapse from the drag loads from ground resistance and the application of brakes, and from the side loads from gusts and crosswinds.

In many cases the undercarriage needs to fit into a very small space, and the units may be turned, twisted or folded to allow this. Retracting an

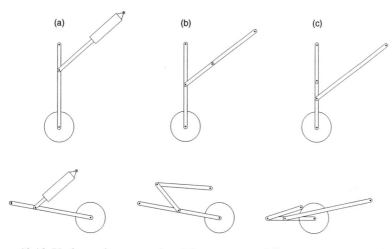

Figure 10.13 Undercarriage retraction. *There are many different arrangements of undercarriage retraction, since the space available into which the undercarriage is to be folded may be restricted in volume or shape by adjacent structure. This diagram shows a few examples of the many options available to the designer.*

undercarriage can be likened to trying to fit a quart into a pint pot.

If the hydraulic system fails, the undercarriage must be capable of being lowered and locked so that the aircraft can land safely. Nose legs usually retract forwards, which means that they will fall under gravity and aerodynamic drag if the retraction system fails. Retraction and extension may be hydraulic, pneumatic or electrical, with their associated benefits and drawbacks, but by far the majority of aircraft use hydraulic systems. It is common for pressure bottles to be fitted, which store enough pressure to allow the undercarriage to be extended once if the system fails.

All retracting undercarriages are required to be positively locked in both positions, and to have lock indications for each leg in the cockpit – the *three greens*. The uplocks and downlocks operate automatically as part of the retraction and extension sequence, usually by a spring and cam or pin to engage them and by a small hydraulic actuator to release them. An independent release, usually by a cable, provides for manual operation. It is imperative that the locks are carefully designed so that they are as reliable as possible, since a malfunctioning lock may cause the undercarriage to collapse. This requires careful attention to materials selection, tolerancing and possible failure modes. In particular, it is important to isolate the down locks from ground loads.

Where an aircraft is used for short-haul feederliner flights, between about 100 and 200 km, for instance, the main-wheel undercarriage doors may be omitted to save weight, and the wheel arranged to lie flush with the outer skin when retracted. In most applications, however, the undercarriage well is closed off by a door or set of doors to reduce drag. The doors to the undercarriage well may be attached to the legs, or may use separate jacks to open and close them. It is important that the doors open before the undercarriage units extend or retract, and usually close afterwards, so a *sequencer valve* is used to direct the supply of hydraulic fluid to operate jacks in the correct sequence.

BRAKES AND TYRES

Brakes

The high weights and speeds at which many aircraft land mean that the braking system must be capable of absorbing and dissipating the very large amounts of heat created as the energy of motion is converted. There are two main types of brake – drum brakes and disc brakes. The drum brake is rarely used, because it suffers from poor heat dissipation, causing the brakes to overheat and *fade* – lose their braking effectiveness at high temperature.

Disc brakes (Figure 10.14) are much more effective at dispersing the heat produced, maintaining their effectiveness during long periods of heavy braking. They consist of a disc or series of discs of aluminium alloy, steel,

Figure 10.14 Disc brake. *The advantages of the disc brake, of light weight and good heat dissipation, make it the only option for all but the smallest of aircraft. The disc brake works by clamping a set of brake pads onto one or more discs. The pads are attached to the undercarriage leg and the discs rotate with the wheel, so the friction between the two slows the aircraft. Because of the enormous energy absorbed, the operating temperatures are very high.*

carbon or other material, gripped between pads of friction material. These pads are forced against the discs by pistons under hydraulic pressure. Control is usually achieved by placing a toe pedal for each brake on its respective rudder pedal. These can then be operated *differentially* by the pilot, giving the ability to steer the aircraft by applying different amounts of braking on each main wheel. Applying the brakes equally on both main units causes the aircraft to be braked smoothly in a straight line. Large aircraft may have quite a number of discs in each brake, to get the required braking forces and heat dissipation. The combination of carbon discs and carbon pads, known as the carbon–carbon brake, is widely favoured, because it combines light weight and the ability to operate effectively at extremely high temperatures.

Braking effectiveness is reduced if a wheel locks (stops rotating), perhaps leading to loss of control of the aircraft. In some instances, the wheel may lock but the tyre may continue to rotate. If this occurs for even a few seconds, the heat generated will cause a tyre burst, or possibly a fire. Locking of the wheels is prevented by an anti-skid unit detecting when the wheel or wheels on any unit stop turning, and momentarily releasing brake pressure on that unit only. This gives the aircraft the ability to stop in the shortest possible distance without loss of control, particularly when surfaces are wet or icy. The unit works in a similar way to the *anti-lock braking system* (ABS) fitted to many cars.

If the wheel and tyre overheat, the pressures generated may cause tyre bursts. Excessive tyre pressure may be prevented by a fusible plug fitted in the wheel. This melts before the tyre becomes hot enough to blow out, and allows the tyre to deflate slowly, easing the steering problems caused by the tyre bursting.

The braking of an aircraft can be supplemented by other forms of braking, which are not part of the undercarriage system. However, it is worth mentioning these systems briefly here, since they are relevant to the complete braking system of the aircraft. Used both in flight and during the landing, air brakes (Figure 10.15) consist of large plates fitted to the fuselage or wings that can be moved so they project into the air flow. They cause a large increase in drag to slow the aircraft. After touch-down, reverse thrust of jet engines can be deployed by moving doors into the jet exhaust to deflect the flow forwards. Turbo-prop engines can achieve a similar effect by changing the pitch of the propeller to reverse the air flow.

Tyres

Tyres are the only contact between the aircraft and the ground (it is hoped!), so they are vital to all aspects of ground movement. The operator is interested not only in their performance, but also their life, since every landing causes tyre wear. The wheels and tyres are accelerated (spun up) by friction as the aircraft touches down, and the friction scrubs off some of the tread. The tyres on some large military aircraft have a very short life. This would be unacceptable to any civilian operator, not only because of the cost but also the time taken to replace tyres frequently. They expect a life of several hundred landings before the tyre is worn out, and it is rare for tyres to fail because of bursting on impact.

Attempts to overcome spin-up wear by powered rotation of the wheels before touchdown have not been widely adopted, because of the extra weight and complexity involved. The enormous weight of many aircraft requires tyres to be inflated to much higher pressures than car tyres – up to 8 bar. The pressure must be checked frequently, because over-inflation dramatically shortens tyre life.

Two forms of tyre construction are available – cross-ply and radial. The cross-ply tyre contains a number of layers of reinforcing fibres, arranged at various angles to support loads in different directions and give the tyres the required stiffness. The radial tyre contains a series of belts under the tread area, with radial cords around the complete tyre profile. The result is a tyre that runs cooler and has lower rolling resistance, but sidewall deflection is greater, which adversely affects control on the ground. For this reason, cross-ply tyres are more common than radial tyres on aircraft.

One or more wire beads around the inside radius of the tyre maintain the diameter, and ensure that the tyre grips the wheel rim effectively. A series of grooves (the tread) is cut on the contact surface of the tyre, to disperse water and allow the tyre to maintain contact with the ground on wet surfaces. Some tyres have *chines* on the sidewalls, which deflect dispersed water sideways, reducing water ingress into engine intakes during the ground run.

Figure 10.15 Air brakes. *Although not part of the undercarriage, air brakes, as seen on the top of the fuselage here, are used in flight and during the landing to increase aerodynamic drag and slow the aircraft. On landing, they supplement the wheel brakes and reverse thrust.* Photograph: Alistair Copeland.

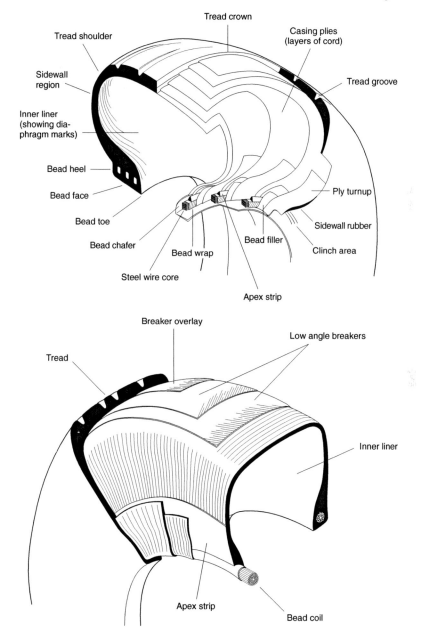

Figure 10.16 Tyre construction. *Aircraft tyres may be of cross-ply or radial-ply construction. Both contain a number of reinforcing plies to maintain the shape of the tyre and contain the high pressures at which they operate. A wire bead or beads around the inside radius of the tyre ensures good contact with the wheel rim to prevent tyre rotation on the rim. The tread disperses water to maintain good contact on wet surfaces.* Illustrations courtesy Dunlop Ltd, Aircraft Tyres Division.

CHAPTER 11

FLIGHT CONTROLS

Objectives: to describe the operation of the pilot's basic controls, and state the reasons for and operation of powered controls.

INTRODUCTION

As described in Chapter Seven, an aircraft must have a control system that allows the pilot to manoeuvre in the three major axes. In addition, the speed of the aircraft must also be controlled. The standard method of achieving this control is to provide a set of control surfaces for pitch, roll and yaw, allowing control of the direction of travel. Some form of control for engine thrust or power is also provided to control the aircraft speed. Although there are variations, the four primary controls are:

- elevators or foreplanes for control in pitch
- rudder for control in yaw
- ailerons for control in roll
- throttle(s) for control of speed

Each control surface is located near the extremities of the aircraft, so utilising the largest moment arm about the centre of gravity, and thus allowing surfaces to be small in size.

In fighter aircraft, computer control can be used to control an unstable aircraft, which would otherwise be impossible to fly. Unstable aircraft controlled in this way are very responsive, providing the manoeuvrability needed to gain an advantage over an opponent.

With the increasing use of computers in aviation, most control circuits, except in the smallest of aircraft, use computers in some way, either to assist or to control the aircraft entirely, by processing demands from the pilot. The autopilots in modern transport aircraft are becoming increasingly sophisticated, and some are now capable of flying the aircraft throughout the whole flight, including take-off and landing. However, many aircraft have some

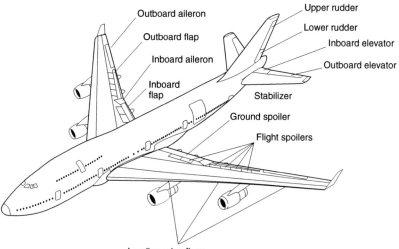

Outboard aileron
Outboard flap
Inboard aileron
Inboard flap
Upper rudder
Lower rudder
Inboard elevator
Outboard elevator
Stabilizer
Ground spoiler
Flight spoilers
Leading edge flaps

Figure 11.1 Commercial aircraft control surfaces. *The Boeing 747 has an extremely complex set of control surfaces, which are linked by the onboard computer systems to improve passenger comfort and reduce pilot workload. Many of the primary surfaces are split, providing fail-safety so that control is maintained if a component or part of a system fails.* Illustration provided by the Boeing Company's Maintenance Training Organization.

form of mechanical back-up in case the main system fails. In the remainder, the control system must be designed in such a way that there is an extremely small possibility of total failure.

CONTROLS

The control column (often known simply as the stick) controls the aircraft in pitch and roll. In large transport aircraft, the stick may be replaced by a yoke (see Figure 11.2), or a side-stick device may be used, for example in the Airbus cockpit. The rudder bar provides control in yaw. Movement of the control column and rudder bar is transmitted to the respective control surfaces by cables or push-pull rods in small aircraft. In large or fast aircraft, the forces required to move the controls are too high for the pilot to operate them, and a system of power operation or power assistance is required, using powered actuators. Throttle control is always via a set of throttle levers, one per engine. An aircraft with a single crew member, or with two crew members in tandem, will have the throttles positioned to the pilot's left. Aircraft with a side-by-side crew arrangement, common to most transport aircraft, will have the throttles between the two crew seats, so that they are accessible to both crew members.

Figure 11.2 Commercial aircraft cockpit controls. *This photograph, from a Hawker 1000, shows the yoke arrangement for pitch and roll control. The additional buttons on the yoke are for controlling other functions, like the radio. The throttle levers, one per engine, are between the two crew seats.* Photograph courtesy Raytheon Corporate Jets Inc.

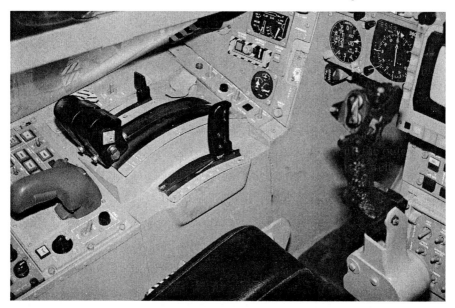

Figure 11.3 Combat aircraft cockpit controls. *The arrangement shown in this Tornado mock-up, of a single stick for pitch and roll control, is used in combat aircraft of all types, and also in many small aircraft. Its advantage lies in its small size and simplicity, although it is then made more complex by the addition of the controls for various systems in the aircraft, to which the pilot needs ready access. The throttle levers and wing-sweep control lever can be seen to the pilot's left.*

Although the way in which the controls are operated varies from one aircraft to another, a given movement of the controls will always make the aircraft turn in the same direction:

- Moving the control column forwards makes the aircraft pitch nose down; backwards lifts the nose.
- Moving or turning the controls to the left raises the left aileron and lowers the right one, making the aircraft roll to the left.
- Pushing the right rudder pedal deflects the rudder to the right, causing the aircraft to yaw right.

STICK FORCES

As aircraft speeds increase, control-surface loads increase as the square of the airspeed. At medium speeds, it is still quite possible to move the controls, but long flights can be tiring. At higher speeds, particularly with larger aircraft, a point is reached where the pilot is physically unable to make the required control movements.

There are a number of solutions to these problems, and the method selected will depend on the magnitude of the control loads. As has already been seen, the centre of gravity (CG) may vary with the amount and distribution of payload, and as fuel is consumed during flight. This CG movement will require a compensating movement of the control surfaces, particularly in pitch, to maintain steady flight. This would require the pilot to hold a steady control input, which would be tiring during anything but the shortest flight. Small deflections of the controls can be maintained using trimming tabs, or trim tabs, which alter the aerodynamic balance of the control surfaces so that they naturally take up a slightly different position relative to the main surface. This has the same effect as a small, continuous load applied by the pilot to the controls, and can be used to set the aircraft to hold steady, straight and level flight, or indeed any other desired trim setting. The trim controls are usually provided as a handwheel or pair of electrical switches for each control – elevators, ailerons and rudder – and are independent of the main control system. In aircraft with fly-by-wire systems, a purely mechanical trim system may be the only form of manual control available in the event of complete system failure.

AERODYNAMIC AND STATIC AIDS

However, trim tabs overcome only one part of the problem. As stick forces increase, the pilot will need some form of assistance to allow him to manoeuvre the aircraft. If stick forces are not too high, it may be enough to provide some simple means of making control movements easier by balancing the control surfaces both statically and aerodynamically. Mass balancing involves ensuring that the mass of the control surfaces is equally distributed about the hinge line, so that the surface does not tend to droop under its own weight or inertia. This would otherwise impose a continuous load on the controls during steady flight, and a widely varying one during manoeuvres. In most control surfaces the majority of the surface is behind the hinge line, so extra mass is added forward of the hinge.

Mass balances and horn balances

The control surfaces will be mass balanced if the centre of gravity is coincident with the hinge line. During the 1950s and 1960s, when high stick forces were becoming more common, this was often achieved by placing a weight on a stalk projecting forward of the control surface, but this is rarely seen in modern aircraft. The usual method is for part of the control surface (the horn) to project forwards of the hinge line, and for a balance weight to be incorporated inside the horn. This also provides some aerodynamic balancing in many cases, since it also moves the centre of pressure forwards. Part of the control surface now operates in a 'scissor' fashion, as can be seen in Figure 11.4, and it is

Figure 11.4 Shrouded horn. *Shrouded horns allow a reduction in stick forces by providing mass- and aerodynamic balance about the hinge line, but the shroud reduces the risk of jamming of the controls by ice ejected upstream during de-icing.*

possible that pieces of ice shed from the wings may become trapped in the gap created and jam the controls. This can be avoided by using a shrouded horn, as seen in the figure. Horn balances are rarely necessary on ailerons.

Inset hinges

Moving the hinge line back can also help with balancing, since it will align more closely with the centre of gravity of the control surface. So the mass required to achieve balance will be reduced, or might not be required at all. As with the horn balance, a greatly inset hinge line can give some aerodynamic balance as well.

Aerodynamic aids use the effects of the airflow around control surfaces to reduce stick forces, thereby helping the pilot to move the controls. As with other aerodynamic surfaces, the distributed load across the control surface can be considered to act at a single point, the centre of pressure. If the centre of pressure is some distance from the hinge line, as is the case with a simple control surface, then the moment produced will require an equal moment from the pilot's controls, and large stick forces will result. Using inset hinges to bring the hinge line back towards the centre of pressure reduces the moment.

Balance tabs

When a control surface is deflected, the centre of pressure moves, and a torque is generated on the entire flying surface. Using this principle another, smaller,

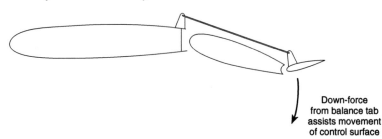

Down-force
from balance tab
assists movement
of control surface

Figure 11.5 Balance tab. *The balance tab operates by generating a large moment on the rear of the control surface, reducing the stick force required to move the surface. A fixed link between the tab and the main surface automatically moves the tab in the opposite direction to the control surface.*

control surface (a balance tab) can be added to the first surface (Figure 11.5). As this balance tab is moved, the resulting torque moves the main control surface in the opposite direction. By linking the balance tab to the control surface in such a way that it moves in the opposite direction, an aerodynamic aid can be created. Taking the idea further, the linkages from the pilot's controls can be connected directly to the tab, for instance moving an elevator tab *up* if the aim is to deflect the elevator *down*. This arrangement is known as a servo tab, but is now rarely used, since most large and high-speed aircraft now use powered flying controls.

Whatever methods are used to reduce stick forces, it is important to recognise that some stick force must be retained so that the pilot receives some tactile feedback in response to control inputs, known as *feel*. If for example a hinge were inset too much, it would be said to be overbalanced, and the surface would tend to deflect without any input to the control column or rudder bar. This would be unacceptable to the pilot, since the stability of the aircraft would be seriously impaired.

POWERED CONTROLS

There are two main types of powered flying controls – power-*assisted* and power-*operated* control systems. These may be used in combination, with, for example, power-assisted controls being used for the rudder and elevators and manual (unassisted) controls for the ailerons, which are generally easier to move. These various ways of moving control surfaces have advantages and disadvantages.

For large aircraft and many fast jets, the use of power-operated controls is the only feasible choice, because the controls would be so difficult for the pilot to move, and because of the control circuits used. For light aircraft and small commercial aircraft the simplicity, lightness and cost of manual controls may be used to best advantage, and no power assistance is necessary.

For aircraft between the two extremes, power assistance is often the ideal solution, with some of the control effort being supplied by the pilot with some assistance from mechanical actuators. This takes advantage of smaller actuators, gives pilots more *feel*, since they are still operating the control surfaces directly, and gives a manual backup if the power-assistance system fails.

Where the control system of an aircraft is totally dependent on maintaining the power supply, alternative systems must be provided in case of failure. If the system is power assisted, and it is possible to fly the aircraft using a manual system in an emergency, then this may provide the required back-up. This is known as *manual reversion*. If it occurred without warning, manual reversion could cause the pilot to lose control due to the sudden increase in stick forces. To avoid this, many aircraft have a hydraulic accumulator, which gives a supply for a small number of control movements whilst the pilot is slowing down and trimming the aircraft in preparation for flying on full manual control. If manual reversion is not adequate because of the stick forces involved, the usual alternative is to provide three or even four independent hydraulic circuits. Each circuit provides power to some part of the control system such that the aircraft can still be fully controlled if one system fails, and can still be flown in an adequate (if limited) way if two systems fail. Other functions that are less vital may be sacrificed or inhibited if this happens, because the vital task of maintaining control in an emergency takes top priority.

COMPUTERS AND FLY-BY-WIRE

Connecting the pilot's controls to the aerodynamic surfaces electrically rather than mechanically is commonly known as *fly-by-wire*, and the surfaces are moved by electric or hydraulic actuators.

Another variant that is becoming available is *fly-by-light*, where the control signals are transmitted by fibre optics rather than electrical cables.

Modern digital computers make it possible to fly an unstable aircraft. The speed of the control computers is high enough to make the constant corrections required as a result of the aircraft instability, giving the pilot the impression of a stable but highly responsive fighter aircraft. The extreme responsiveness comes about because any manoeuvres are assisted by the aircraft's instability. For a fighter aircraft, this can present an enormous advantage in combat.

For civil transport, extreme control responsiveness and unstable aircraft are obviously not required, but computers and fly-by-wire systems are still useful, because their rapid information processing increases safety by prohibiting control movements that would create a hazard to the aircraft. Since all control demands are processed by the computer, rather than fed directly to the controls, any control movements that could endanger the aircraft, for instance by inducing a stall, can be ignored. This creates an aircraft that is almost impossible to stall, which is the cause of many aircraft accidents.

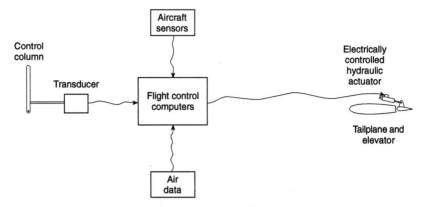

Figure 11.6 Fly-by-wire-system. *Fly-by-wire systems operate by processing movements of the controls using a computer. The computer generates the required outputs and sends them as electrical signals to the actuators. The actuators then provide the necessary motion of the control surfaces. The resulting system is lighter than alternative systems, and provides easy integration with the autopilot and related systems.*

Computers can break down, and software is notoriously difficult to free of errors, so all computers that are essential to the control of an aircraft must be triplicated. They operate on a voting system, where if one computer disagrees with the other two it is outvoted, the majority decision carries and a computer-failure warning is given. Most fly-by-wire aircraft employ four independent control systems.

In many aircraft, especially the older ones, computers are used only for the autopilot, and the outputs are used to move the aircraft controls by servo motors attached to the standard control cables or linkages. In this way the autopilot duplicates the actions of the pilot directly, and the pilot can see and feel the control movements generated. There is more information on autopilots in Chapter 12.

ACTUATORS

For powered flying control systems, there is a range of possible actuators to choose from. The most common actuator for operating control surfaces, as well as many other moving parts, is the hydraulic actuator. This type of actuator provides high jack forces from a small unit. Hydraulic circuits operate at around 200 bar (3000 pounds per square inch). Because of its design, a hydraulic jack can create higher forces in extension (pushing) than retraction (pulling). A detailed description of the function of a jack can be found in Chapter 13.

ENGINE CONTROLS

The primary control for the engines is concerned with controlling their power output or thrust. Other controls may include, for piston engines: propeller pitch, fuel mixture, carburettor anti-icing, magneto selectors; for gas-turbine engines: reverse thrust, re-heat, propeller speed (constant-speed turbo-props).

There is a fundamental difference between the way that engine output is controlled in piston engines and in gas turbines. In piston engines, throttling is achieved by controlling the amount of air flowing into the engine. The carburettor is responsible for ensuring that the fuel-to-air ratio is maintained at the correct value, with some limited provision for adjustment available to the pilot. A fuel-rich mixture gives more engine power, and is used for take-off and climb out; a fuel-lean mixture gives less power, but is more economical, and is used for cruise conditions. The pilot's throttle lever is connected directly to the carburettor, in the same manner as that in a car, and an additional lever is provided to control fuel mixture.

In gas-turbine engines, no direct control of air flow is possible, and engine output is controlled by adjusting the fuel supply to the engine. If the fuel flow is increased, the fuel-to-air ratio will be increased, higher temperatures and pressures will be generated inside the combustion chamber, and the turbine speed will increase. Because the turbine or turbines drive the compressors, more air will be drawn in, and the overall mass flow in the engine will increase, generating more thrust (or power in the case of a turbo-prop). As the air flow increases, the fuel-to-air ratio will fall, and equilibrium will once again be achieved, but at a higher thrust than before. As may be expected, reducing the fuel flow will have the opposite effect. The rate at which the fuel flow is increased determines the rate at which the engine thrust changes, but if it is increased too quickly, the pressure in the combustion chamber becomes too high, and the smooth flow through the engine is disturbed. The flow will reverse back through the compressor, a condition known as a *surge*. Serious damage can be done to the compressor, and the combustion flame may be extinguished, causing temporary loss of engine thrust, or even complete engine failure. So it is important that the rate of increase of fuel flow is limited under all flight conditions.

Because it is not feasible to control the rate of movement of the throttle levers, especially in an emergency, it is not sufficient to connect the throttle levers directly to a device that adjusts fuel flow in accordance with the lever position. The connection goes instead to a fuel-control unit, which monitors engine conditions, and adjusts fuel flow as required and at the ideal rate to match the demands of the pilot. In large transport aircraft, this has been taken a stage further, with a system known as FADEC (Full-Authority Digital Engine Control). With a FADEC system, the throttle levers are calibrated to demand a given thrust level rather than fuel flow rate. The FADEC system receives these demands and adjusts the fuel flow to provide the thrust demanded. With all gas turbine engines there is a relationship between thrust and engine speed (usually

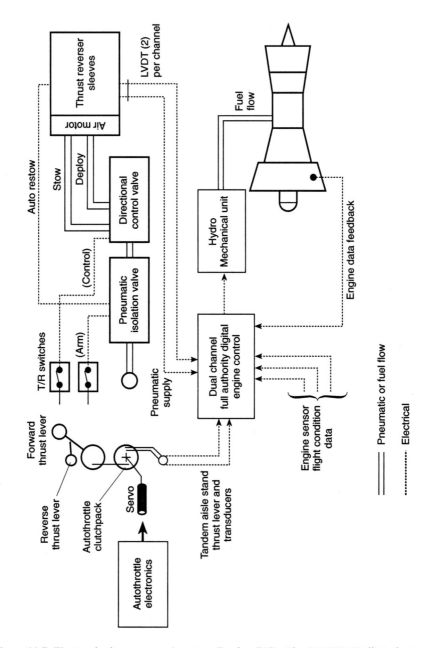

Figure 11.7 Electronic thrust control system (Boeing 747). *The FADEC (Full-Authority Digital Engine Control) system takes inputs from the pilot's throttle levers, engine sensors and flight data and calculates the exact fuel flow rate to provide the required thrust. The pilot has no direct control over the fuel flow, hence the control system is said to have full authority.* Illustration provided by the Boeing Company's Maintenance Training Organization.

expressed in terms of percentage of the nominal maximum rpm), but it is also dependent on altitude and temperature. So if the pilot demands 100 per cent of the rated thrust, and this can be achieved using only 90 per cent rpm, FADEC provides these settings. However, if the required thrust cannot be achieved without exceeding the engine rpm limits, the thrust provided will be limited automatically. Thus the FADEC system has full authority over the engine controls, as its name infers, which conserves engine life and provides more consistent control over the engines, reducing pilot workload.

OTHER FORMS OF CONTROLS

So far, we have considered an aircraft of conventional layout. For aircraft of different layouts, most of the previous descriptions still apply, but in certain respects different control surfaces will be required.

Elevons

Tailless aircraft (i.e. without a tailplane) do not have conventional elevators, so the functions of conventional ailerons and elevators are combined into elevons. Elevons can be deflected together to act as elevators, or differentially to behave as ailerons. The two forms of deflection may be mixed to achieve combined pitch and roll actions simultaneously, of course.

Canard controls

This type of control surface (shown in Figure 11.8) usually replaces the conventional tailplane, and is often used together with a delta wing. Canard

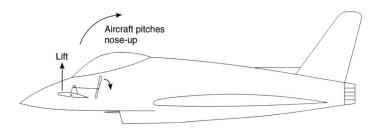

Figure 11.8 Canard controls. *Typical of the arrangement used for many delta-winged aircraft, canard controls can give enhanced manoeuvrability when combined with fly-by-wire in an artificially stable aircraft. The benefit gained by this makes it a popular arrangement for the new generation of agile combat aircraft. An increase in angle of attack of the canard foreplane gives an increase in lift, pitching the aircraft nose up – the opposite effect to increasing the angle of attack on an all-flying tailplane. The control column movement, of course, remains the same with both types of aircraft; pulling the column back lifts the nose.*

control surfaces are usually of the all-flying type, rather than separate deflecting surfaces, which gives improved protection against stall problems. Increasing the angle of attack of the foreplane increases the lift, lifting the nose and causing the aircraft to pitch up, which is in the opposite sense to the effect of elevators. The control column still operates in the same way, of course, so the pilot sees no difference in the control movements required to fly the aircraft. There are a few aircraft that have both canard and conventional control surfaces, and pitch control may be provided either by elevators or by canard controls.

The all-flying tail

At high speeds, the camber change caused by deflecting the elevators becomes less important than the effective change of angle of attack, so the elevator becomes less effective. It is advantageous to change the angle of attack of the entire tailplane in this case. Although a tailplane and elevators are structurally simpler, the all-flying tailplane (Figure 11.9) is effective at all speeds and Mach numbers, so all high-speed aircraft of conventional layout use this method.

The variable-incidence tailplane

An alternative to using trim elevator tabs is provided by using a small actuator under the leading edge of the tailplane to change its angle of incidence for pitch trimming, but retaining the conventional elevators for pitch control. At

Figure 11.9 All-flying tail. *At high speeds, elevators become less effective, so the all-flying tail becomes a better alternative. Hydraulic actuators change the angle of attack in response to movements of the control column.*

high subsonic speeds this method is more effective than trim tabs, but retains the structural simplicity of conventional elevators.

Tailerons

Tailerons (Figure 11.10) are similar in concept to elevons, but in this case the two halves of an all-flying tailplane are moved independently to do the job of ailerons, and together to act as elevators. This allows flaps to be fitted across the entire trailing edge of the wing, because separate ailerons are no longer needed. On some aircraft, the tailerons are supplemented by spoilers on the wings, which are operated differentially.

Spoilers as ailerons

Normal ailerons cause the aircraft to roll by increasing lift on the wing that is to rise and reducing the lift on the wing that is to descend. However, it is quite possible to roll the aircraft using just one of these actions, although the effect will of course be reduced. Deflecting an aileron down to increase its lift also increases the drag on that wing, which will tend to cause a yaw in the opposite direction to the yaw induced by the roll, and so would be undesirable. But raising the aileron on the down-going wing alone will increase drag on that wing, assisting the turn. Rather than using one aileron to achieve this, spoilers

Figure 11.10 Tailerons. *Operating on the same principle as elevons, tailerons combine the functions of elevators and ailerons by moving both together and differentially. They may be supplemented by spoilers, also operating differentially. Using tailerons allows flaps to be fitted across the entire wing trailing edge, which improves the landing performance of aircraft with high wing loading, such as this F-16.* Photograph: Alistair Copeland.

can be operated only on the wing that is to descend. Some aircraft employ a complex combination of aileron and spoiler deflections to achieve roll control, selecting the optimum combination for each specific flight condition.

Inboard and outboard ailerons

Because lift increases as the square of the speed, the effects of camber change, such as deflecting control surfaces, increase likewise. So the effectiveness of ailerons increases with air speed. For transport aircraft, this is rarely desirable, since the most extreme control responses are needed at low speed, to offset the effects of turbulence during the approach, for instance. It is imperative that the ailerons are capable of correcting large turbulence and gust effects, so that the aircraft can be adequately controlled during this phase of flight, but such controls may be too sensitive for flight at high speed. Normal ailerons are placed near the wing tips to maximise their effectiveness, so smaller ailerons placed further inboard will have reduced effectiveness – ideal for high-speed flight (see Figure 11.1). Software settings in the flight-control system will provide a gradual change-over from the low-speed conventional ailerons to high-speed inboard ailerons as aircraft speed increases.

Rudder ratio changers

The same reasoning that is applied to inboard and outboard ailerons can be applied to the rudder. Since there is no practical method of reducing the moment arm of the rudder, many modern transport aircraft have a system that changes the rudder ratio, i.e. the rudder deflection for a given movement of the rudder pedals. Large deflections may be required at low aircraft speeds to correct for any gusts or crosswinds, especially during the final stages of landing, or to correct for thrust asymmetry caused by an engine failure. The ratio change is performed automatically by the flight control system, reducing rudder deflections at high speed, and may be achieved electronically or mechanically, depending on the control system used. Another method that can be used is to have a two-part rudder, with upper and lower sections. One section is used at high speed, and both sections at low speed. This has the added advantage of fail safety – if the actuators on one of the sections fail, the other section can operate independently, although of course the effectiveness of the rudder at low speeds will be reduced.

Pitch control by CG movement

An aircraft can usually be trimmed to a steady attitude in roll and yaw without requiring significant control trim deflections and therefore high drag. This is not always the case for pitch trim, because it is heavily influenced by the aircraft centre-of-gravity position. If the CG moves aft, it will generate a nose-up pitch trim change, which must be balanced by a nose-down moment from the elevators or tailplane, whichever is used for trimming in pitch. By moving mass fore and aft, the CG position can be controlled in flight, removing the need for this trimming and its attendant drag. Developing this further, the pitch attitude of the aircraft could be changed by moving the centre of gravity position. It would be inconvenient to move the payload around, but a large portion of the mass of most aircraft is given over to fuel. By providing an additional fuel tank in the tail, a long way from the centre of gravity, pitch trim may be changed by a relatively small movement of fuel between this tank and the main tanks. This allows pitch *control* without the drag penalty associated with aerodynamic control-surface movements. Some large aircraft have this facility, with the autopilot using this method unless large pitch demands are made.

CHAPTER 12

AUTOPILOT, RADAR AND RELATED SYSTEMS

Objectives: to describe how an autopilot works, and how it can be extended to fly the aircraft through a wide range of manoeuvres; describe a range of related equipment, such as radar, terrain-following, instrument landing systems and auto-land; and describe how these systems work and how they can help to ease the pilot's work load in flight.

INTRODUCTION

Many flights in modern aircraft are of long duration, often several hours. Even in fighter aircraft, a large number of Combat Air Patrol (CAP) sorties may be flown, where the aircraft stays airborne for a number of hours. To make the job of a pilot as simple as possible, routine flying on a given heading can easily be performed by a mechanical or electronic system. This system is called an autopilot. With the use of computers, a much wider range of flying can be carried out by the autopilot, and it is possible to make a complete flight with little or no pilot input. This relieves the pilot of routine tasks, which can cause fatigue, and leaves him free to concentrate on other aspects of the flight.

PRINCIPLES OF OPERATION

If the autopilot is to be of practical use, it must be able to keep the aircraft in straight and level flight, or in steady climbs and descents, when required. It would also be very useful if it could carry out simple turns when a change of heading is needed. This is much more difficult, because it needs a correct mixture of the three main controls – elevators, ailerons and rudder.

Modern autopilots are often much more sophisticated than this, and can carry out complex manoeuvres such as landings safely and reliably. When an F-14 fighter takes off from the deck of an aircraft carrier, by catapult, the autopilot system controls the take-off, with the pilot taking control only after the aircraft is safely airborne.

This section describes how the simplest forms of autopilot work. To control an aircraft requires two main operations:

- a way of detecting when the aircraft has strayed from the required heading and attitude
- a system for calculating what control movements are needed to correct this error, and for making those control movements

Using this system, if the flight path of the aircraft is disturbed, perhaps by a *gust* (a sudden change in wind direction), a control movement is produced to restore the desired attitude. This control movement needs to be proportional to the size of the disturbance, so that the appropriate correction is made. Because the aircraft needs to be controlled in three axes – pitch, roll and yaw – three *channels* are needed in the autopilot, one to control in each axis.

The device that detects the disturbance is a *rate gyroscope*. Conventionally, this is an electrically driven spinning mass. The mechanical gyroscope tends to maintain a fixed orientation in space, so any change of aircraft attitude will cause the gyroscope to turn or attempt to turn relative to the aircraft. This motion can be detected electrically by a *pick-off*. An electrical signal is thus generated which depends on the *rate* or speed of this disturbance. The signal is then amplified, so that it can be fed to the second part of the system, the *corrector* circuit.

In our basic autopilot, the disturbance correctors are *servo motors* that are designed so that the speed of movement is proportional to the size of the signal applied to them. Because this signal comes from the detector, the corrective output will be proportional to the rate of the disturbance. As the aircraft returns to its original attitude, the error signal will reduce, and the correcting action will be reduced accordingly. The servo motors operate on the control cables or linkages in the normal flight control system in small aircraft, and even on some larger aircraft. On more complex aircraft, electronic circuitry replaces the servo motors, and the entire system is incorporated into the electronics and software of the flight-control computers.

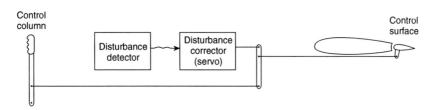

Figure 12.1 Principle of autopilot operation. *The basic autopilot consists of a device for detecting when the aircraft has deviated from the required course, and a servo motor or actuator for supplying the required corrective control movement. When the aircraft has returned to the required course, the error signal becomes zero, and the control surfaces are returned to their normal positions. Each channel of the system detects and corrects disturbances in one axis, so three channels are required for full control.*

Apart from the basic spinning-mass type gyroscope, there are many other types, which use vibration or light to measure the disturbance. In particular, the ring-laser gyroscope (Figure 12.2) is becoming more common, because it is more accurate and much more reliable than a mechanical gyroscope. It consists of a triangular or square ring of glass, which is hollow. Two beams of laser light travel around the ring in opposite directions, reflected at the corners. If the gyroscope is rotated, one beam will take slightly longer to travel around the ring than the other. This tiny time difference can be detected, and an electrical signal can be produced in proportion to the speed of rotation.

The simple mechanism of controlling an aircraft's attitude described so far has some limitations, which need to be overcome. Because of the way in which the autopilot works, the aircraft will gradually move away from the heading and attitude set in the autopilot, which is known as *drift*. This requires another independent system that the autopilot system can use to check for gradual

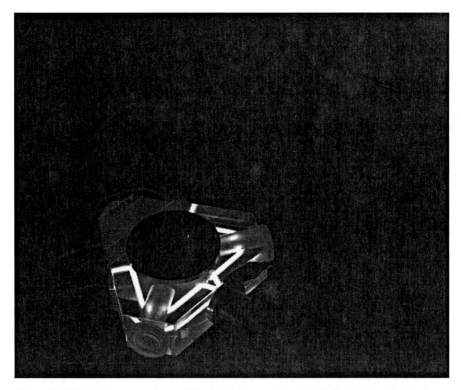

Figure 12.2 Ring-laser gyroscope. *The ring-laser gyroscope (RLG) detects rotation by passing two beams of laser light in opposite directions around a closed tube. Because the speed of light is constant, any rotation of the gyroscope will change the length of the path of the two beams. The changed path length will create a small difference in the time taken for each beam to travel around the ring. This time difference can be measured, and used to generate a signal that is proportional to the speed and direction of rotation.* Photograph courtesy British Aerospace (Systems & Equipment) Ltd – BASE.

changes and correct them. Pitch and roll drift can be detected by a reference that detects which direction is vertically down, similar to a pendulum, and heading drift can be detected using the compass as a reference.

An aircraft heading could be controlled using just the rudder, but it is better to use ailerons in conjunction with the rudder, or even to use them *instead* of the rudder, rolling the aircraft until it turns to the required heading. So the autopilot can be set up to send correction signals from the *yaw detector* to the *aileron corrector*. This is known as *cross-feed*, and results in smoother corrections of yaw disturbances.

MANOEUVRES USING THE AUTOPILOT

So far, the autopilot has been considered only as a means of maintaining a set heading and attitude. The error detectors generate control signals that are proportional to any error between the aircraft attitude and a reference. If this reference is deliberately changed, however, the autopilot will interpret the difference between the aircraft attitude and the new reference as an error, and deliver the required signals to bring the aircraft attitude to the new setting. Early autopilots achieved this by mounting the error detectors on a stabilised platform, which incorporated a gimbal system – a system of pivots arranged so that the platform could be rotated to any attitude relative to the aircraft. By deliberately moving the platform, the aircraft could be made to stabilise in any desired attitude, still under autopilot control. By gradually rotating the platform horizontally, for instance, the aircraft would smoothly turn to a new heading.

Modern systems are much more sophisticated than this basic system. The autopilot can be used to control entire phases of the aircraft's flight by linking the system to a navigation system. The navigation system detects the geographical position of the aircraft, using navigation beacons and often satellite navigation systems, and aircraft altitude using barometric or radar altimeters. Linking the autopilot to instrument-landing, inertial-reference and other systems, a complete, automatic flight-control system can be produced.

Naturally, the autopilot system itself is rather more sophisticated than the simple model described here, and the majority of the functions are now represented in software in the aircraft's computer systems, but the principles remain.

YAW DAMPERS

Modern airliners are required to be extremely stable in flight, for the comfort and confidence of passengers, and for safety. There are forms of aerodynamic instability that can be difficult to design out in aircraft with highly swept wings, one of which is the Dutch roll. Barnard and Philpott[1] present a detailed

[1] RH Barnard and DR Philpott, *Aircraft Flight*, Longman

description of this form of instability and its effects. In essence, it is a combination of rolling and yawing which are out of phase, and can be extremely unpleasant to passengers. Many airliners reduce the effects of Dutch roll artificially by using yaw dampers (Figure 12.3) as part of the flight control system. In the process they provide better co-ordination of rudder and aileron movement to improve turns and reduce the effects of turbulence. Information from the inertial reference units, air data computers and accelerometers is used to develop an accurate indication of the motion of the aircraft, and corrective outputs are sent to the system controlling the rudder. Any tendency for a Dutch roll to begin is detected and the relevant corrections applied before the motion develops.

RADAR SYSTEMS

Aircraft fly at speeds and altitudes that require better information than can be gleaned using the unaided eyes of the pilot. Aircraft at distances of only a few miles are difficult to see, and the high speeds at which aircraft fly mean that potential collisions leave little time for safe avoiding action to be taken. What is needed is some way of improving the limited vision that nature has provided. Radar (RAdio Detection And Ranging) can provide this help, in a number of forms.

The basis for any radar system is the reflection of radio waves by solid objects; a narrow beam swept in a specific pattern will give the position of such an object, usually in azimuth (bearing) and elevation, from the transmitter. Using a pulsed radar, and timing the return of the reflected pulse, allows the distance from the transmitter to be calculated.

As would be expected, the systems fitted to civil and military aircraft are radically different, although some systems are common to both types.

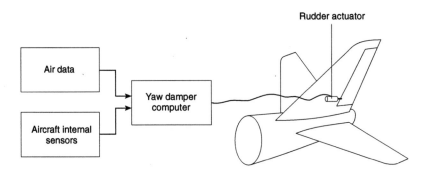

Figure 12.3 Yaw-damper system. *The yaw damper is an artificial way of preventing Dutch-roll instability by taking data from various sources to identify the onset and provide the necessary control inputs to damp out the instability.*

Civil aircraft systems

Anti-collision radar supplements the procedures and systems used by air traffic control to provide an extra level of safety. High terrain and other aircraft are displayed, and warnings may be provided if these hazards are within a predetermined distance. Inputs from the radar altimeter may also be integrated into this system, to provide ground-proximity warning (see below).

Weather radar provides information on weather conditions in the area ahead of the aircraft. It is widely used in civil transport aircraft, since bad weather, particularly lightning and hail, can be a hazard. The turbulence associated with storms can also reduce structural life, and cause discomfort and worry to passengers. Reduced visibility in heavy cloud is also a factor, although not a major one. Armed with weather information, the crew will normally divert around storms. Weather information is displayed on the navigation display screen (Figure 12.4).

TCAS (Traffic-alert and Collision Avoidance System) detects other TCAS-equipped aircraft in the vicinity and predicts potential collisions, providing warnings and collision-avoidance guidance to the crew.

Radio altimeters are fitted on many aircraft, since a significant cause of aircraft accidents is collision with high ground, particularly when associated with navigation errors. Radar altimeters accurately measure the true height of the aircraft above ground, as opposed to the barometric unit which is normally used to provide altitude information in terms of sea-level altitudes. Actual height information is of prime importance at low level, such as during approach; altitude above sea level is the normal measure in cruise flight, since it uses a standard reference.

Ground-proximity warning is derived from a number of sources, the prime one being the radio altimeter. If the aircraft is too low, two stages of warning are provided – initially, a caution, comprising visual and audible signals or voice, is used. If the aircraft is lower still, a warning signal, now comprising klaxons or 'pull up' voice signals and flashing warning lights, must convey the urgency without generating panic on the flight deck.

Since aircraft must approach the ground to land, it would be undesirable if these events occurred each time the aircraft landed, so inputs from various sources, including undercarriage position, stall warning, flap position and information on the stage of flight, may be used to inhibit the system or to provide various warnings or cautions beyond a simple ground-proximity warning. Examples of this can be seen in the Boeing 747-400, which warns of 'excessive wind shear', 'excessive descent rate', 'excessive terrain closure rate', 'excessive descent after take-off or go-around', 'unsafe terrain clearance when not in landing configuration' and 'below glide slope'. It also provides voice call-out of radio altimeter height information during landing. All of these warnings can be inhibited if required, so that emergency situations are not complicated by spurious warnings and the crew can concentrate on priority actions.

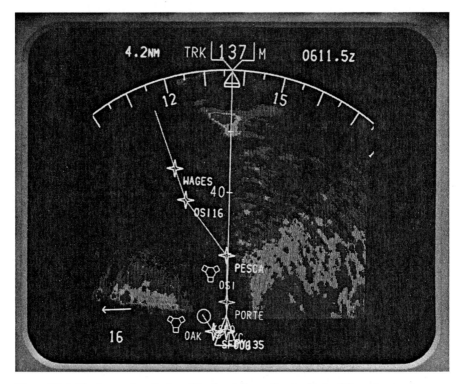

Figure 12.4 Weather-radar display. *Weather information is obtained from the weather-radar system, installed in the nose of the aircraft. The information allows diversion around poor weather conditions, reducing the risk of structural damage, lightning strike and extreme turbulence. This photograph shows the display from the weather radar (the shaded areas) together with navigation data.* Photograph courtesy Collins Commercial Avionics.

Military aircraft systems

Air-intercept (AI) radar is the primary radar system on combat (fighter) aircraft. It detects possible targets within a given range of the aircraft, and often supplies information to missile systems, so that they can be programmed with target information before launch. Using pulse-Doppler radar allows range to be calculated by timing the return of each pulse, and range rate (the speed of the target towards or away from the aircraft) using the Doppler shift of the echo from the original radar pulse. Combining this with the bearing and bearing rate (obtained by comparing bearing information from successive pulses), a complete picture of the target's position and speed in three dimensions can be gained.

    Terrain-following systems use a special radar, more complex than a simple radio altimeter, which maps the ground in front of ground-attack aircraft (Figure 12.5). The system then anticipates the control movements needed to fly the aircraft close to the ground without colliding with it. As with the auto-land

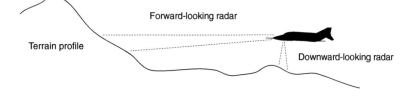

Figure 12.5 Terrain following. *Using terrain-following radar, an accurate picture of the shape of the terrain ahead allows an aircraft to fly safely at high speed and low altitude. When this system is linked with an autopilot system, the pilot can set the height above ground at which the aircraft is to fly, and the accuracy with which the height is to be maintained. This allows the aircraft to avoid detection as it approaches ground targets.*

system, the autopilot is used to control the movements of the control surfaces. In this way, a ground-attack aircraft can fly at high speed at very low level, giving the minimum chance of being detected by enemy radar. The system is so fast that the aircraft flies at heights and speeds beyond the capabilities of an unaided pilot.

The system can be set to control the aircraft height as precisely as required, within limits, so the pilot can choose whether the aircraft is to follow the ground profile accurately to minimise detection, or less accurately, which provides a smoother and slightly more comfortable ride.

RELATED SYSTEMS

A wide range of systems may be fitted to modern aircraft, many of which provide additional functions either to supplement those provided by the autopilot or to allow the autopilot itself to offer improved functionality. Some of these systems have already been mentioned, but a more complete description is provided here.

Satellite navigation/global position systems (GPS)

There is a constellation of 24 *GPS* satellites placed in orbit by the US military, for both civil and military use, although the signals are configured so that civilian users cannot achieve the same accuracy in use. A similar system, called Glonas, has been placed by the Russians, and pilots around the world have access to both systems. Satellite-navigation receivers vary greatly in both size and sophistication, from simple hand-held sets used by private pilots to highly complex units fitted to many aircraft. Basically, the principle for all of these units is the same. The satellites send out continuous time-coded signals which are decoded by the receiver on the aircraft. The network is such that at least three satellites are in direct line of sight at any time, and by analysing the signals from three or more satellites the distance from each of them may be

Figure 12.6 Satellite-navigation/GPS. *Using transmissions from satellites, the aircraft's position in space (i.e. in three dimensions) can be determined to within a few metres anywhere in the world. The equipment to receive this information varies from quite cheap, hand-held sets shown here, to comprehensive systems incorporated in the navigation suite in both military and civil aircraft.* Photograph courtesy of Trimble Navigation.

determined. The satellites follow the same ground track on each orbit, and so the exact position of each satellite is known at a given time. Triangulating the position from the satellites allows the position of the receiver, and hence the aircraft, to be determined in three dimensions – latitude, longitude and height. Using this system, the position of the aircraft can be determined anywhere in the world within close limits, typically fifty metres for civil users and one or two metres for military users.

Automatic direction finding (ADF)

Many areas of the world operate non-directional radio beacons (NDBs), which broadcast a signal continuously at a frequency which is specific to that beacon. Thus any aircraft receiving the signal can identify the station. An aircraft equipped with ADF equipment analyses the signal received from the beacon and displays the bearing to the beacon, relative to the heading of the aircraft, on the ADF display. Adding the aircraft heading to this relative heading gives the actual bearing to the beacon. Using at least two ADF bearings, the position of an aircraft can be found by plotting the intersection. Some ADF stations also transmit weather information.

VOR produces a similar result to ADF, but works slightly differently. VOR stations transmit an identification signal followed by a direction signal which is swept in a circle. The VOR system on the aircraft decodes this swept signal and determines the bearing of the *aircraft* relative to the *VOR* station (rather than the other way round, as in ADF). Unlike the ADF system, variations in the aircraft heading do not affect the bearing displayed on a VOR. A course deviation indicator allows the aircraft to be flown along a line radiating from the VOR station (a radial), displaying deviations from the radial so that course corrections may be made.

Distance-measuring equipment (DME)

Most VOR stations also transmit a DME signal, which can be used to derive the range of the aircraft from the beacon. No height information is included, so the range displayed is the actual distance – the hypotenuse of the triangle given by the height of the aircraft and its ground distance from the beacon. However, this is acceptable for most purposes.

ADF, VOR and DME displays are normally incorporated into the horizontal situation indicator or navigation display (see Chapter 15).

TACAN is a military implementation of a combined VOR and DME system.

Instrument landing systems (ILS)

These use a signal from a transmitter close to the runway to give an indication to the pilot when the aircraft is on the correct approach path (glide slope), both horizontally and vertically (Figure 12.7). This is useful when visibility is poor and the runway cannot be seen. ILS is not part of the autopilot system; it just gives the information to the pilot on a special instrument, or more commonly on a multi-function display (see Chapter 15).

Auto-land systems process information about the aircraft's height, position, heading and speed relative to the runway, using information from the ILS system in addition to aircraft data. They then generate instructions to the autopilot to fly the aircraft on a suitable heading to achieve a touchdown at the correct point on the runway. These systems are effectively fully automated versions of the ILS described in the previous paragraph, and are highly accurate and a useful safety feature. Automated take-off is also possible, using a pre-programmed take-off routine. Another safety feature is the 'go-around'. By pressing a single button, a landing can be aborted (for instance if there is an obstruction on the runway) and the aircraft will automatically climb away to re-join the circuit.

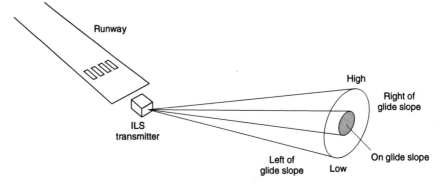

Figure 12.7 Instrument landing system. *A radio beam from a location close to the touch-down point gives an indication on a cockpit display to help the pilot fly 'down the beam' when visibility is poor. The system works by displaying, on a cockpit instrument, any deviation from the correct bearing and glide slope. Thus the pilot can manoeuvre the aircraft until it is correctly aligned with the touch-down point.*

Fully programmed flight

It is now quite possible to fly the aircraft automatically, by programming way-points, arrival times, flight levels, etc. into the autopilot system. The aircraft system will set the speed, height and bearing as required, making automatic corrections for wind speed and direction, and allowing the flight plan to be modified at any time by the pilot. By combining some or all of these systems, an aircraft could fly an entire flight without pilot intervention. In the future, it is likely that many civil aircraft will not have pilots at all, but will be pre-programmed to fly a particular route unaided. System failure is largely prevented by having several back-up systems, and as a last resort the aircraft could be controlled from the ground to complete a flight safely. The main problem with this would not be a technical one; it would be passenger resistance to a pilotless aircraft. However, once it can be shown that automatic systems are safer than human pilots, this means of flying will certainly become available.

CHAPTER 13

AIRCRAFT SYSTEMS

Objectives: to describe the following systems and how they work: hydraulic systems, pneumatic systems, electrical systems, fuel systems, de-icing and anti-icing systems, auxiliary power units, communications systems.

INTRODUCTION

Besides the flight-control and autopilot systems, an aircraft contains a vast array of other systems. The aircraft must be completely self-contained in flight, and power of various kinds is required at many locations within the aircraft to perform a wide range of tasks.

Other than in the smallest and most basic of aircraft, the movement of many controls and systems, such as control surfaces, undercarriage and flaps, cannot be performed by direct mechanical linkages from the cockpit. Some form of power operation is required, and the choice of power source depends very much on the type of system being operated. Each method of creating motion has advantages and disadvantages, and some are better suited to certain tasks than others. Most aircraft carry a variety of power systems for this reason, although there is considerable cross-linking of systems for safety. This chapter describes the features and operation of the major systems to be found on most aircraft.

HYDRAULIC SYSTEMS

Hydraulic systems are a convenient way of providing the force required to retract and extend undercarriages, to operate flaps and wheel brakes and many other movements. They are also widely used to move control surfaces, such as ailerons and elevators, in power-assisted and power-operated control systems. Although the pumps and pipework needed to supply pressurised hydraulic fluid around the aircraft might seem to be quite heavy, the system can be much lighter than an electrical or mechanical system. This is because of the high pressures used, which produce small units of high power. Hydraulics have a distinct advantage over some other systems: because hydraulic fluid is almost incompressible, very precise control of the position of actuators may be achieved, making it ideal for use in powering flight controls.

The basic hydraulic system consists of one or more engine-driven pumps, drawing fluid from a reservoir and pumping it into a system of pipes, which distribute fluid around the aircraft. Another set of pipes returns used fluid to the reservoir. Since hydraulics may not be in use all the time, hydraulic accumulators are usually incorporated in the system, which store fluid under pressure for use during peak demands. Pumps driven by the aircraft engines may be fixed or variable delivery. Fixed-delivery pumps operate continuously, and a pressure-relief valve returns excess fluid to the reservoir when demand is low. Variable-delivery pumps supply only what is required, and so are more efficient, taking power on demand. (It is important to recognise that pumps deliver a flow *rate*, not pressure – the pressure in the system occurs as a result of resistance in the system to the fluid delivered.) In the event of failure of the aircraft engines or pumps, electrically driven pumps can be brought into use, and emergency accumulators may also be provided for short-term use if all power is lost. Most aircraft have several hydraulic systems, normally three or four, arranged in such a way that if any system fails the others are able to maintain all of the systems with little or no degradation in function. Each of

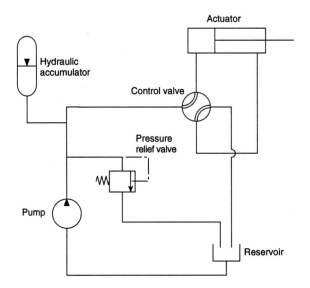

Figure 13.1 Basic hydraulic system. *A basic system requires a pump, to pressurise fluid drawn from the reservoir and supply the fluid into the pressure line. Control valves divert the pressurised fluid to the required side of hydraulic jacks to cause them to extend or retract. Used fluid flows back to the reservoir via the return line, completing the circuit. Accumulators store pressurised fluid and smooth out the pressure in the system. The pressure-relief valve limits the pressure in the system, returning excess fluid to the reservoir when demand is low.*

the flying controls is operated by at least two jacks, operating on different circuits, and each is capable of operating the controls independently.

Hydraulic systems use hydraulic fluid, a mineral- or synthetic-based oil, to transmit pressure. The system is set to operate at a pressure of about 2100 newtons per square centimetre (200 bar, 3000 pounds per square inch). This means that a hydraulic jack capable of exerting a force of one tonne (10 000 newtons) would only be about 25 mm in diameter. Hydraulic jacks or actuators come in a variety of shapes and sizes, but the basic principle of operation is the same. To extend the jack, oil is passed into one end of the jack, and the other end is opened to return, allowing the fluid to pass unrestricted into the pipes which return the fluid to the reservoir. The resulting pressure difference, acting on the surface area of the piston face, creates a force which causes the jack to extend, and at the same time ejects the fluid from the return side of the jack. To retract the jack the connections are reversed, and pressurised fluid passes into the other side of the jack, where it acts on the area on that face of the piston. Note that jacks generate less force when retracting, since the effective area of the piston is reduced by the cross-sectional area of the piston rod. By controlling the rate at which the oil flows into the jack, the speed of its extension or retraction can be controlled.

A jack will normally have two connections to it – one to each side of the piston. A selector valve switches the fluid pressure between the two lines to control operation of the jack, with the non-pressurised line returning ejected fluid. Of course, the jack need not necessarily move through its full travel each time it is operated, allowing precise control over its position.

Where the actuator is required to move to a specific position, a servo actuator is normally used (Figure 13.3). A servo actuator will follow the control linkage automatically. This type of actuator is especially suited to power-assisted controls – where the control linkage is also connected to the control surface being moved, the actuator helping the pilot to move the control

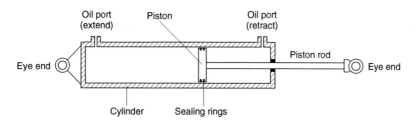

Figure 13.2 Hydraulic jack or actuator – operation. *The hydraulic jack works by generating a force as a result of the pressure difference across the piston. Since the non-pressurised side of the piston is vented to the return line, its pressure is close to atmospheric pressure, so the force produced is equal to the system pressure multiplied by the effective piston area. The effective area of the piston for retraction is reduced by the cross-sectional area of the piston rod. Provided the maximum available jack force is not exceeded, the jack will continue to move until it either reaches the end of its travel or the supply of pressurised fluid ceases.*

surface – and also to power-operated controls.

Actuators do not represent the only form of motive power that can be generated using hydraulics, although they are by far the most widely used. Hydraulic motors, operating like a pump in reverse, can also be used to provide motion at relatively low speeds but with high torque. Operating through a gear box, a hydraulic motor can be used to drive an electrical generator, another example of the cross-linking of systems for safety. Hydraulic circuits can be cross-linked using a combined hydraulic motor and pump, together called a *power-transfer unit*, which can supply one circuit from another without the risk of fluid loss from both systems in the event of failure in either system.

An aircraft hydraulic system is quite complicated, but is made up of a lot of very simple circuits. The system is similar in many ways to a set of electrical circuits, although of course a flow of oil is being used rather than a flow of electricity. The use of valves to control hydraulics means that these valves can be made to operate automatically. Using *sequencing valves*, a series of operations can be carried out just by making a single selection. For instance, retracting an undercarriage requires a series of movements that must be carried out in the correct order, and at the correct time. Selecting 'undercarriage up' will begin a sequence of operations, starting with opening the undercarriage doors, then unlocking the undercarriage downlocks, lifting the legs into their bays, locking the legs in the stowed position, then closing the undercarriage doors. The presence of hydraulic pressure at various points in the system, and/or the positions of jacks and locks, can be used to inform the crew when the undercarriage is locked and unlocked, by means of a set of indicator lights.

Hydraulic fluid, being a type of oil, presents a fire risk, especially when it is at high pressure, because any small hole will cause a fine mist of oil to be produced. Hydraulic fluids are usually inhibited, to reduce their flammability. The near-incompressibility of hydraulic oil is an important feature of hydraulic systems, and it is important that all traces of air and water are kept out of the system or the precision of movement will be lost.

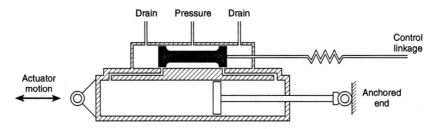

Figure 13.3 Servo actuator – operation. *A servo actuator is a variant of the standard actuator, and incorporates a servo valve. The servo valve controls the jack, and is operated by a linkage set up so that the jack will follow the motion of the linkage. When the jack has reached the position set by the linkage, the servo valve will automatically close off the supply and the jack motion will stop. This type of actuator is common in powered flying controls.*

PNEUMATIC SYSTEMS

Pneumatic systems work in a very similar way to hydraulic systems. The difference is that in pneumatic systems high-pressure air is used instead of hydraulic fluid. Like hydraulics, pneumatic pressure can be stored in a accumulator. This can give a reserve of power for short bursts of very heavy operation, or for emergency use if the system fails. Using air as a medium for transmitting motion has another advantage – no return line is needed. The exhausted air is vented directly to atmosphere, although the venting of large volumes of air under pressure must be allowed for in the design of nearby structure.

However, the compressibility of air can be a major disadvantage. Pneumatic systems lack the instant response that hydraulic systems provide, and the rate of movement of pneumatic actuators is highly load-dependent. This compressibility also means that the position of systems needing partial movements, such as control surfaces, cannot easily be controlled with any degree of accuracy, since even when the flow has stopped the actuator will move in response to load variations.

Another disadvantage of pneumatic systems is their inefficiency in transmitting power, because energy is lost in compressing the air. This does not occur with hydraulic fluid.

Because of these major disadvantages, many aircraft are not fitted with a pneumatic system. However, many aircraft use *compressor bleed air* to do certain tasks. Because gas-turbine engines generate hot air at relatively high pressure, a small amount of this air can be used in anti-icing, as will be described later in this chapter. Bleed air is used on many aircraft to supply the power and heat to operate cabin pressurisation and air conditioning systems. This air is not passed directly into the cabin, but through a *cold-air unit* (Figure 13.4), where it is cooled and then mixed with the correct amount of hot air to provide the optimum temperature for the comfort of passengers and crew.

The humidity is also important, and this is controlled by adding the required amount of water, as determined by a device called a *humidistat*.

ELECTRICAL SYSTEMS

Most modern aircraft have airborne systems which require a large amount of electrical power to operate. These systems include radio, radar, aircraft instrumentation, operation of some secondary controls such as wing flaps, and engine starting. With fly-by-wire systems, the demand on electrical supplies is even greater. Electrical power is also widely used to control hydraulic and pneumatic systems and their associated indicators.

It would be impossible to store the amount of electrical energy needed to supply all of these systems for an entire flight, so the electrical supplies needed are generated during flight. A combat aircraft electrical system will typically consist of two generators or alternators, driven by the aircraft engine or

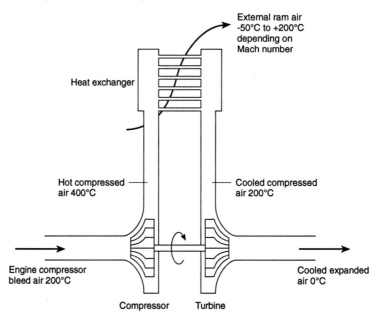

Figure 13.4 Cold-air unit. *The cold-air unit (CAU) works rather like a gas-turbine engine, but energy is extracted rather than added. Air from the gas-turbine compressor passes through the CAU compressor, where its temperature and pressure are raised even higher. The air then loses heat in a heat exchanger. The cooled air then passes through the turbine, giving up energy to compress the incoming air, further cooling in the process. The air, now at very low temperature and pressure, is mixed with hot air to supply temperature-controlled air to the cabin.*

engines through a suitable gearbox. DC generators produce direct-current electrical power, normally at 28 volts, whereas alternators generally produce AC three-phase power (see below). Most alternators must be driven at a constant speed regardless of the engine speed, to provide AC power at the correct frequency, and this is achieved by using a *constant-speed drive unit*. Using two or more generators or alternators ensures that the supply will be maintained even if an engine fails. Multi-engine aircraft, such as many civil airliners, and twin-engined aircraft with wing-mounted engines, will normally have an electrical generator or alternator drive on each engine. Because the output of alternators is predominantly three-phase AC, the output of each is carefully kept in phase with the others.

Only a relatively small amount of energy can usually be stored in batteries, sufficient to start the engines or auxiliary power unit, or for running essential supplies for a short time in an emergency. These batteries are recharged during normal operation. Emergency and auxiliary power provision is described later in this chapter.

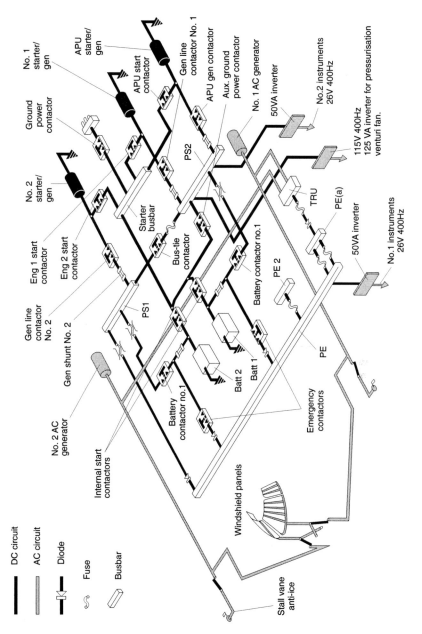

Figure 13.5 Aircraft electrical system. *The electrical system in most aircraft is quite complex, with parallel redundant systems for safety. All essential systems can be supplied from more than one of these systems, and the systems themselves can be cross-fed if required. In this way no single failure, and only a limited number of unlikely multiple failures, can cause a complete loss of electrical power.* Illustration courtesy Raytheon Corporate Jets Inc.

Except for the smallest of aircraft, the electrical power supplies are standard on all aircraft. Two supplies are provided, alternating current and direct current:

- **115/200 volt AC, 400 Hz, three-phase**. Three-phase power is supplied at 115 V line, corresponding to 200 V between phases (see Figure 13.6). Power can be drawn from a single phase if required, provided the overall consumption per phase for the entire aircraft is kept reasonably balanced. Three-phase power is used for most systems in the aircraft that require medium or large amounts of electrical power, because the high voltage means that current is low for a given power utilisation.
- **28 volt DC**. This is used for most electronic systems, provided they do not require large amounts of power. It is also used for cockpit instrumentation, because its low voltage makes it much safer. DC generators can be combined with electric starters (starter-generators) on some engines, reducing overall weight and complexity.

Many aircraft systems, particularly radar and navigation equipment, need electrical supplies at different voltages and frequencies to these. However, from the basic supplies it is straightforward to convert them to the needs of particular equipment. The conversion is usually carried out inside the item of equipment requiring it.

If required, main DC power can be converted to AC using static inverters, and AC converted to DC by transformer-rectifiers, again giving protection against failure of individual systems. This is an important safety feature, further enhanced by the use of hydraulically driven generators, the auxiliary power unit (described later) and the batteries. So under almost any conceivable

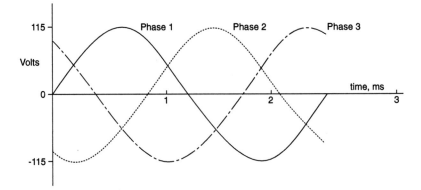

Figure 13.6 Three-phase electrical supply. *The three-phase supply consists of three separate voltage supplies, each phased at 120° to the others. The potential of each line is at 115 V peak with respect to aircraft ground (earth), and 200 V peak with respect to the other phases. So a three-phase supply contains three live lines and one ground. Since the three lines are out of phase, each line acts as a return to the others at some part of the cycle, or 'ground' may be used, depending on how the equipment is wired.*

combination of failures adequate supplies will be maintained until the aircraft can be brought safely down. Many of these features can be seen in Figure 13.5.

When the aircraft is on the ground, it is not always convenient to run the engines or auxiliary power unit every time power is required, and the battery supply may not be adequate, so a socket is provided in a convenient location for a ground supply to be connected from a portable source.

FUEL SYSTEMS

An aircraft in flight uses a large amount of fuel, perhaps as much as 50 000 litres per hour for a very large aircraft at take-off thrust. So a large volume of fuel must be stored in the airframe if the aircraft is to be capable of a reasonable flying time. Most is stored in the wings, but many aircraft also store fuel in the fuselage, and sometimes in the tail (Figure 13.7). The fuel is stored in sealed compartments, or flexible fuel bags may be incorporated where sealing of the structure is not possible. Military aircraft frequently make use of external drop tanks (Figure 13.8); these are used to carry large extra volumes of fuel to give improved range or endurance, but can be jettisoned in an emergency or for combat, so they do not restrict the aircraft performance.

As the fuel is used, the centre of gravity of the aircraft will change, so the fuel will often need to be moved between tanks to keep the CG within acceptable limits, or the aircraft stability could be adversely affected. To allow this movement of fuel, and of course to supply the fuel to the engines as required, a system of pumps and valves is fitted. To prevent the fuel boiling off at high altitudes where the air pressure is low, the fuel tanks are pressurised.

Most aircraft will usually be refuelled through a single refuelling point, although large aircraft may use several points to reduce turn-round time. From this point the fuel management system distributes the fuel as required to the various tanks in the aircraft. For some military missions, the amount of fuel an aircraft can carry is not enough to provide the endurance or range required. Air-to-air refuelling (Figure 13.9), from a specially adapted tanker aircraft, can be used to supply the extra fuel needed. The aircraft will rendezvous with the tanker at a pre-determined point, and the tanker will trail a refuelling hose. A probe on the aircraft connects to this hose, and fuel is transferred.

DE-ICING AND ANTI-ICING SYSTEMS

As discussed in Chapter Six, under certain conditions of temperature and humidity, ice will form on the external surfaces of the aircraft. In flight, it will be concentrated in the areas where pressure and temperature fall as a result of the local air flow conditions – wing and tail leading edges, engine intakes, propeller leading edges and carburettor venturis. This icing can create a hazard for various reasons:

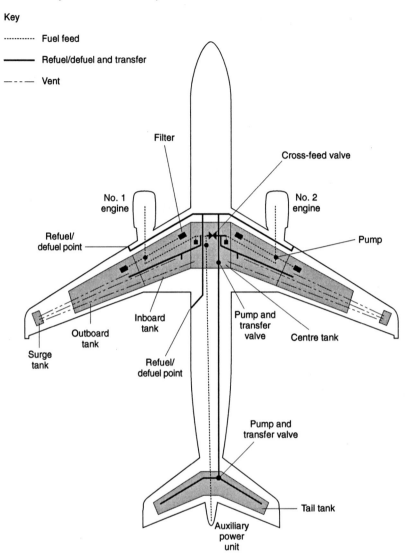

Key

·········· Fuel feed

─────── Refuel/defuel and transfer

─ ─ ─ Vent

Filter

Cross-feed valve

No. 1 engine

No. 2 engine

Refuel/defuel point

Pump

Inboard tank

Pump and transfer valve

Outboard tank

Centre tank

Surge tank

Refuel/defuel point

Pump and transfer valve

Tail tank

Auxiliary power unit

Figure 13.7 Aircraft fuel system. *This is a greatly simplified illustration of a typical commercial-aircraft fuel system. The system must be capable of transferring fuel into and out of each tank, during refuelling and flight, and must also be capable of supplying each engine from either or both sides of the system, for safety. This is achieved using a complex array of pipes, pumps and valves. The tail tank not only increases fuel capacity, but also allows greater control of centre of gravity in flight, to trim out the tail load.*

Figure 13.8 Drop tanks. *The use of drop tanks on combat aircraft allows long-range operations to be carried out, but in an emergency or for combat they can be jettisoned so that aircraft performance is not adversely affected. Connection to the aircraft fuel system is by self-sealing valves to prevent loss of fuel after jettison.*

Figure 13.9 In-flight refuelling. *The useful range of a military aircraft can be considerably increased by using in-flight refuelling. The tanker trails a hose, and the refuelling probe of the aircraft is manoeuvred until it connects with the hose; fuel can then be transferred from the tanker.* Photograph: Sgt Rick Brewell (RAF). British Crown Copyright/MOD.

- On leading edges, the build-up of ice can be considerable, increasing the weight of the aircraft and thus increasing stalling speed, and also disturbing the aerodynamic shape.
- On engine intakes and propellers: again the shape can be disturbed but, more importantly, pieces of ice breaking away from these surfaces can cause damage to structure adjacent to the propeller disc, since they can be ejected at high speed, and gas-turbine compressor and fan blades are susceptible to ice damage by ingestion of intake ice.
- Carburettor icing can quickly cause reduction of power or even complete engine failure.

Icing is of such concern to aircraft designers and operators that many aircraft have lights fitted which illuminate the leading edge of the wing at night, so that any ice accumulation can be readily seen. External probes may be used to detect the formation of ice on external surfaces. Icing of fuel is also a problem – fuel always contains a small amount of water, and if ice crystals form in the fuel they can rapidly block fuel filters and may cause engine failure. To prevent this, the fuel may be heated using a fuel/oil heat exchanger, transferring the unwanted heat in the engine oil into the fuel.

There are two methods of reducing or eliminating the problem of external icing – the ice can be prevented from forming (anti-icing), or it can be removed at intervals which are sufficiently short to prevent significant build-up (de-icing). The method chosen depends in each case on a number of factors.

Anti-icing

Anti-icing may be achieved by electrical heating, heating using hot air or chemically by leaking fluids which depress the freezing point of the water. The chemicals flow through a large number of tiny holes in the surface, and spread to cover the most vulnerable areas. Propellers, especially in twin or multiple configurations where they are located alongside the fuselage, need anti-icing, because even small pieces of ice ejected from propellers could cause damage to nearby structure. They are most conveniently protected by electrical heating, as are carburettors. Wing and tail surfaces and engine air intakes may also be fitted with anti-icing systems, operating either by electrical heating, or more commonly by directing air taken from the compressors of the main engines, which may be at 300°C, onto the inside of the leading-edge skin (Figure 13.10).

De-icing

De-icing is very commonly used for preventing ice build-up on wing and tail leading edges, because it is economical to operate, and effective. A rubber boot, conforming to the profile of the aerodynamic section, is fitted as part of the leading edge of the surface to be de-iced (Figure 13.11). Every few seconds it is inflated, using compressor bleed air or from the pneumatic system, breaking

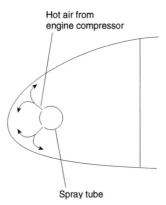

Figure 13.10 Anti-icing using compressor bleed air. *Ice may be prevented from forming on leading edges and engine intakes by heating the skin on its inside face using hot air bled from the gas turbine compressor. A series of pipes distributes the air as required, and small holes in the pipes direct air onto the skin.*

up any ice that has formed and allowing it to be removed by the air stream. This would not be suitable where small ice fragments could damage components or structure downstream, but is ideal for most wing and tail surfaces. Some de-icing systems are electrically operated, with heaters switched on periodically to cause accumulated ice to weaken and break away. As with most anti-icing systems, de-icing is switched off when icing conditions no longer persist.

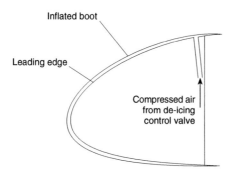

Figure 13.11 De-icing using inflatable boots. *Ice may be removed before it can build up to hazardous levels by inflating a boot fitted to leading edges. This system is more economical in use than anti-icing, because it does not expend energy in preventing ice forming. It is not suitable where the fragments of ice could cause damage to equipment downstream.*

AUXILIARY POWER UNITS

An aircraft is almost totally dependent on the engines to provide not only thrust for flight but also power to drive hydraulic and electrical systems. Most aircraft use powered flight controls, and if all supplies to these are lost the pilot may not even have the facility to make a safe forced landing. The aircraft will have batteries, but the small amount of energy which they can contain, compared to the enormous power requirements of even a limited number of emergency systems, rules these out as a stand-by power source for flight controls. Batteries are limited to providing the initial power source for engine starting, and to supplying a small amount of emergency power.

It is possible to store energy to power hydraulic systems, as stored pressure in a hydraulic accumulator, but again this has very limited capacity and is normally used to lower flaps and undercarriage, and perhaps to power a very small number of control surface movements.

So what is needed is another engine, which can act as a standby power source if the main engines are shut down or fail. Such a device is a small gas-turbine engine, called the *auxiliary power unit*, or APU. The APU can drive electrical generators and hydraulic and pneumatic pumps, and is also frequently used to provide power, either electrical or air, to start the main engines. It can also supply pressurised air for cabin air conditioning if required. On the ground, during passenger embarkation, the APU is used to supply power for all of the systems used. Because the APU is a small engine, it can be easily started using battery power, therefore batteries can be smaller than if they were required to start a main engine. The output from many APUs is sufficient to allow simultaneous starting of all main engines, saving time on the ramp. It is not normally run in flight, unless a main engine fails, in which case it will provide a second power source for the aircraft systems.

The APU may be located anywhere on the aircraft, but of course it needs a flow of air into it, and provision to disperse the exhaust gases. On many airliners, the APU is located in the aftmost end of the fuselage, and on combat aircraft it is typically located above the main engines aft of the cockpit. Since it is fuelled by standard aviation turbine fuel, it can use fuel from the same tanks as the main engines.

Ram-air turbines

It is extremely rare for all main engines to fail during flight, but one possible cause for this may be a fuel-related problem, either with its supply or quality. In this event, the APU, fuelled from the same source as the main engines, will probably also be lost, and in any case it takes time to start. Some aircraft, both civil and military, have a *ram-air turbine* (RAT), which is deployed into the air stream automatically if power is lost. It acts like a windmill, with the air stream turning blades, which in turn power electrical and/or hydraulic power generators. The RAT can provide control in the first few vital seconds, allowing time for the crew to take action to bring the aircraft down safely.

INFORMATION SYSTEMS

As well as the most obvious form of information transmission, that of voice, a vast amount of electronic data is exchanged in modern aircraft, both within the aircraft and to stations outside.

External voice transmissions are by radio, using a variety of wavebands depending on the category of aircraft – military aircraft have separate bands available to them, for obvious reasons, but may also use civil frequencies, for instance when communicating with air traffic control. The two types of radio in common use are VHF (very high frequency) and HF (high frequency). VHF generally provides better quality transmission, because of the higher frequencies used. HF transmissions can travel much further for a given transmitter power, because the longer wavelength suffers less from atmospheric absorption and is less directional, following the curvature of the earth more readily. So VHF is used for ranges of a few tens or hundreds of miles, and HF is used to allow aircraft to keep in touch with their base wherever they are in the world.

Internal voice transmissions, i.e. intercoms, are required so that the crew of a combat aircraft are kept in constant contact – the environment of a fast jet does not lend itself to normal speaking. In civil aircraft, the flight deck may be quiet enough for the crew to converse without resorting to using an intercom, but they may need to talk to cabin staff or passengers.

However, by far the greatest amount of information is exchanged between individual units of both analogue and digital systems. Modern aircraft contain large amounts of avionics, and each function of each system usually involves gathering or providing information. Until quite recently, each unit would be linked directly with the other units with which it shared data, and with any system other than the most simple one the amount of wiring needed is considerable. One way of overcoming this problem, and allowing much greater flexibility in developing additional functions, is to use a *data bus* (Figure 13.12), which forms a central transmission route for all data. The aircraft has a data bus which runs throughout the aircraft, and all systems are connected to the bus. Information is passed to the bus by each system in turn, and all the data traffic is regulated by the *bus controller*, to a standard protocol. The military communication standard for current aircraft is one developed by the US Department of Defense, MIL-STD-1553, and this is progressively being adopted by almost all military aircraft in the West, and by many civil aircraft as well. Any system requiring information can find it simply by receiving it from the bus at the correct time in the cycle. Of necessity, the data transmission rate is extremely high, and priority is given to the most important information and that requiring the highest data rates. As with most systems on aircraft, fail-safety is an important consideration, and it is normal for the bus to have back-up systems. Many aircraft have duplicated data buses, which are synchronised to safeguard data integrity. Using a data-bus system allows some systems to be incorporated that would be ruled out in practical terms using the earlier methods. Central maintenance computers, which take information from almost every other system on the aircraft, are one example. All system status information, particularly of failed systems and components, is gathered together into one computer, which can link with other systems to provide crew

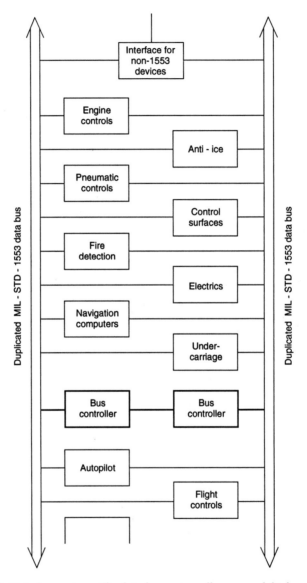

Figure 13.12 Data-bus system. *The data-bus system allows a much higher usage of information by making it available to any device connected to the bus. This vastly reduces the amount of wiring required, and improves flexibility of the functions provided by each system. The heart of the system is the bus controller, which controls which device is allowed to pass information onto the bus at any time. Any number of devices may read data from the bus simultaneously.*

alerts, a print-out for ground engineers, or transmit the information ahead by airborne data link, as required. This provides a safer system, with improved reliability and shorter turn-round times.

Many avionics systems are based on a microprocessor, but do not process data at a constant rate. For much of the time the processor will be idle, but it must have the processing capacity to handle peak demands. Systems designers are now looking towards providing a centralised data-processing system, attached to the bus, which can handle most or all of the data processing for the entire aircraft. Using perhaps four processors for fail-safety, avionics would become smaller and lighter, since much of the redundancy of processors would be eliminated. The power consumption could also be drastically reduced. Since each unit is currently unable to share its processor with other units, if their processor has failed for instance, overall reliability would be improved, and it would be easier to provide better and faster functions from most systems. Each system would carry out the necessary pre-processing of data, then pass it to the centralised processing units via the data bus. After processing, the resulting data is then fed directly onto the data bus, where it can be accessed by any system requiring it.

However, this is for the future, and there are many problems. It requires higher rates of data transfer than current systems, since there is more information to be transmitted, and it increases the already serious problem of software integrity. All software used in aircraft, particularly that associated with flight control systems, must be free of errors. Errors are notoriously difficult to eradicate in software, since the smallest of errors may remain hidden for years, only coming to light when a specific set of conditions occurs. It is impossible to replicate every possible combination of conditions, since the number of inputs may be high, and the resulting error may range from minor to catastrophic. The amount of data code used is increasing exponentially, and this problem may never be satisfactorily overcome.

CHAPTER 14

WEAPON SYSTEMS

Objectives: to describe the range of weapons fitted to military aircraft, the targets they are used against and how a missile operates.

INTRODUCTION

There is a huge range of weapons fitted to modern aircraft, and the choice depends on what job the aircraft is intended to do. The range of weapons can be split into three main categories: fighter weapons, such as air-to-air missiles (AAMs); ground/sea attack weapons, such as guns, bombs, rockets and air-to-surface missiles (ASMs); and electronic/defensive systems, such as electronic countermeasures (ECM) and flares. These divisions are not absolute, because for instance guns are widely used in fighter engagements.

A guided missile may be defined as an unmanned, self-powered, steerable weapon, designed to attack a specific target or range of targets. Similar weapons include guided free-fall bombs and unguided (ballistic) rockets, but neither of these satisfies the definition of a guided missile.

FIGHTER WEAPONS

Air-to-air missiles are the first choice for air-defence (fighter) aircraft. They have a much greater range than guns, and the aircraft does not need to point exactly at the target when the missile is fired. The selection and launch are controlled by the pilot, using the aircraft's weapons-management system. When a target is within range, the details will be relayed to the pilot, usually through the head-up display, and sometimes by sound in the headset. The pilot then selects the type of missile to be fired, using the weapon-selector switch, and fires the missile when the aircraft is in the best position to attack the target.

When the launch signal is sent to the missile, a complex chain of events is set in progress. The launch sequence depends on the type of missile to be fired, but

Figure 14.1 Sky Flash air-to-air missile. *Tornado F3 carries four Sky Flash missiles, to provide its medium-range capability. The missiles are semi-buried within the fuselage, and must be ejected downwards from the aircraft before the rocket motor fires.* Photograph courtesy British Aerospace Dynamics Ltd.

will include firing the special batteries used to supply electrical power to the various parts of the missile, starting gyroscopes in the missile navigation system, telling the missile guidance system where the target is, and perhaps telling the missile what frequency its radar is to operate on. The missile will also self-test all of its electronics to make sure everything is working properly. If all is well, the rocket motor which powers the missile will fire, and the missile will be launched. If the missile fails its self-check it will not be fired (a *hang-fire*), a signal will be sent to the aircraft, and the aircraft system will automatically select and fire another missile.

Guns (cannon) are used as a back-up to guided missiles in air combat, having a much shorter effective range. Their main use is in the air-to-ground role.

GROUND/SEA ATTACK WEAPONS

A ground-attack aircraft will typically carry one or two guns (cannon) of 20–30 mm calibre (barrel internal diameter). They can fire at a very high rate (between 20 and 100 rounds per *second*, depending on the gun), and the ammunition is heavy, so they carry ammunition for perhaps ten seconds of gun firing. They may be used against a variety of targets from armoured vehicles to troops. Some aircraft cannon use special ammunition which is very effective against heavy armour.

Air-to-surface missiles (ASMs) are very specialised weapons, their design depending on the target they are designed to engage. Anti-armour weapons

Figure 14.2 Aircraft cannon. *Cannon are extensively used for ground attack against a wide range of targets. They are particularly effective because of their relatively large calibre (hence high projectile mass) and high rate of fire, but their ammunition supply is limited for the same reason. Cannon may be used as a back-up for air-to-air missiles.* Photograph supplied by British Aerospace Defence Ltd – Royal Ordnance Division.

Figure 14.3 Sea Eagle anti-ship missile. *Using a radar seeker and sea-skimming capability the Sea Eagle, shown here on the Tornado GR1, is capable of engaging ships in all weathers at ranges of more than 50 km. It travels at high speed close to the water (sea skimming), so it is difficult to detect until it is very close to the target, reducing the effectiveness of countermeasures.* Photograph courtesy British Aerospace Dynamics Ltd.

have warheads which are designed to penetrate a considerable thickness of armour plating, even reactive armour (which explodes when attacked to deflect the effects of the warhead). Anti-radar weapons home in on sources of radar transmission, such as surface-to-air missile sites, and defeat them by destroying the antenna using a shower of small fragments (shrapnel). Anti-ship weapons skim close to the surface of the sea, to delay their detection by the target. Many missiles are able to 'see' the target in great detail, and identify a range of possible targets by comparing the seeker image with images stored in a computer memory. These missiles can be programmed to engage certain targets in preference to others, even flying past or over some targets to reach another with a higher priority.

A special kind of ASM is the cruise missile. This is a very long range missile (often over 100 km), which is programmed to navigate and fly using ground mapping and terrain following (in a similar way that a motorist might follow a map), and inertial navigation. Cruise missiles are extremely accurate, striking within a few metres after flying many kilometres. They are not just launched from aircraft – they may be launched from ships or land vehicles as well.

Figure 14.4 Rocket pod. *The rocket pod carries a number of ballistic rockets and can fire them singly or in salvoes. Although unguided, rockets can be extremely effective against a range of ground targets. The pod shown here, with its front cover removed, holds nineteen rockets.*

Rockets are unguided (ballistic) weapons, often fired several at a time (a *salvo*), against ground targets. Because they are unguided, they are most effective at fairly short range, but can be very effective against lightly armoured vehicles and ground installations.

There is a wide range of bombs available, to perform a variety of different tasks. The simple high-explosive type can be used for many purposes, and comes in a range of sizes. A variation of this type is used to penetrate deep into concrete. Retarded bombs (Figure 14.5) have a small parachute which slows the bomb down rapidly. This allows the bomb to be dropped very accurately from low altitude without risk to the aircraft. The aircraft is thus less vulnerable to ground fire.

Laser-guided bombs (Figure 14.6) are a compromise between the effectiveness of a missile and the low cost of a bomb. A standard bomb is modified by the addition of a laser-detection system and a set of steerable fins. The target is illuminated by a laser, either from an aircraft or from a man on the ground, and the bomb steers itself by homing in onto the laser energy reflected by the target. The bomb is free-fall, so its range is highly dependent on its launch altitude, but can be extremely accurate, making possible a technique known as *surgical bombing* – accurate bombing of targets with minimum damage to their surroundings.

Cluster bombs, which open after launch to release a large number of bomblets, are highly effective against armoured and 'soft' ground targets. The

Figure 14.5 Parachute-retarded bomb. *A method of achieving high accuracy when bombing, with reduced risk of the launch aircraft being shot down, is to release bombs from very low level. Parachute retarders are used to slow the bomb rapidly to allow the aircraft to clear the target area before the bomb explodes.* Photograph courtesy Irvin Aerospace Ltd.

Figure 14.6 Laser-guided bomb. *A laser target-designator – fitted to the launch aircraft, a targeting aircraft or operated from the ground – provides a marker onto which the laser-guided bomb can home. Steerable fins direct the bomb onto the target with very high accuracy, as seen in the Gulf War.*

bomblets are dispensed to cover a specific area, and each bomblet has an armour-piercing penetrator and a fragmenting case.

Specialist bombs, such as the JP233 airfield-attack system (Figure 14.7), are also used. The JP233 dispensers contain a large number of small bombs, which are scattered over a runway. These bomblets are of two types. One type creates deep craters in runways, preventing aircraft using the runway. The second type is an area-denial munition, which explodes if lifted or tipped, causing considerable damage to heavy repair vehicles. They also self-destruct at various times, severely hampering repair teams.

ELECTRONIC/DEFENSIVE SYSTEMS

Just as the pilot of each aircraft has defined missions to carry out, there are others whose job it is to thwart the mission, usually by destroying the aircraft. These threats may come from the air or the surface, and combat aircraft are normally provided with some form of countermeasures against such attacks. Of course, several of the systems already described may be used to defend the aircraft, but there are also systems which have no offensive capability – their sole purpose is to protect the aircraft. There are various methods which may be used, but many rely on the simple nature of radar systems and missile seekers.

Figure 14.7 JP233 airfield attack weapon system. *Two JP233 systems can be seen attached to the underside of this Tornado. Each consists of two dispensers, containing 30 cratering munitions and 215 area-denial munitions. The area-denial munitions cause great damage to heavy repair vehicles, severely hampering efforts to repair damaged surfaces. JP233 is normally jettisoned from the aircraft after the munitions have been released.* Photographs courtesy Hunting Engineering Ltd.

Figure 14.8 BL755 cluster bomb. *The cluster bomb is released in the same way as a normal bomb, but then breaks open in flight to release a cloud of bomblets, each of which combines armour-piercing and fragmentation functions. This makes cluster bombs effective against both armoured and 'soft' ground targets.* Photograph courtesy Hunting Engineering Ltd.

One method is so simple it is easy to forget – that of making the aircraft as difficult to see as possible, both visually and on radar. The current favourite term for this is *stealth*, but of course aircraft have had some form of camouflage since the first world war. Against radar, the shape of the aircraft is very important, because the amount of radar energy reflected is determined just as much by its shape as by its size. An active form of camouflage can also be used, where any incident signal is met by some form of counter signal, to make it appear that the aircraft is somewhere else (false echoes) or to attempt to scramble or swamp the signal reflected from the aircraft in a jumble of electronic noise. These systems are known as electronic countermeasures (ECM), and most combat aircraft have an ECM capability in some form.

If this ploy fails, then further defensive action is required, by deflecting or decoying incoming missiles. Some simple infra-red guidance systems may be fooled into following decoys, in the form of pyrotechnic flares dropped from the aircraft. These are more intense than the infra-red transmission from the aircraft itself, but most modern infra-red guided weapons are sufficiently sophisticated to recognise flares and ignore them. Radar-guided missiles might be diverted by clouds of *chaff*, thin strips of aluminium foil which reflect radar signals strongly; the aim is to deflect the attention of the guidance system. Alternatively, they may be fooled by ECM, but some missiles are able to detect the presence of a jamming signal and home in on it instead of using their own radar.

Theoretically, it is possible to destroy an incoming missile with another missile, an anti-missile missile, but these are not yet sufficiently developed for practical use on aircraft. They are, however, successfully used on some ships, and it is perhaps only a matter of time before they become small and capable enough to be used on aircraft.

It may appear that it is almost impossible to evade modern weapon systems, and that may be true in some circumstances, but the chances of detection and interception of any aircraft depend greatly on its shape, the prevailing environment (particularly the electromagnetic environment), and a large degree of pure chance. Cost is also a factor – all of the above systems are expensive, and there can be a cost spiral, where every countermeasure is met by a counter-countermeasure. Each development makes the systems more expensive, and in the real world this means that fewer can be afforded. A balance is therefore needed between capability and quantity, always a difficult decision to make.

HOW A MISSILE SYSTEM OPERATES

The function of a missile system falls into three main tasks:

- locate (and track) the target
- fly to the target (within the warhead kill radius)
- destroy the target, or render it ineffective

Locating and tracking the target

To locate and track the target, the missile system must have some means of detecting and identifying it. The unit which does this is generally called a seeker. It detects some form of emission from the target, in the form of radar emission or reflection, infra-red emission or reflection, optical sighting by an operator, or magnetic influence or acoustical emissions in the case of some torpedoes. Most air-launched missile seekers work by either reflection of radar signals emitted by the missile or launch aircraft, or by detecting infra-red emission from the target.

Radar seekers are of three forms – active, semi-active or passive (Figure 14.9). An active seeker carries its own transmitter, and detects the reflections from the target of the missile's own radar transmission. This makes it extremely capable of withstanding jamming, since the radar characteristics can be varied to prevent the target setting up false radar images. An active seeker gives the missile a *fire-and-forget* capability, allowing the launch aircraft to engage another target immediately, or to depart immediately after launch. This form of seeker is ideally suited for medium-range missiles (say, 15–30 km maximum range).

If greater range is required, it is difficult to obtain sufficiently high transmission power to locate targets. In this case, semi-active seekers can be used. These work in a similar way to active seekers, but the transmitter is

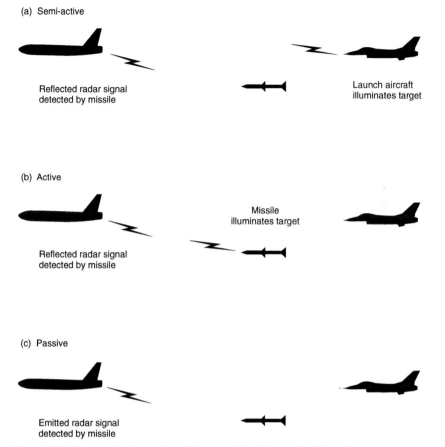

(a) Semi-active

Reflected radar signal
detected by missile

Launch aircraft
illuminates target

(b) Active

Reflected radar signal
detected by missile

Missile
illuminates target

(c) Passive

Emitted radar signal
detected by missile

Figure 14.9 Types of radar seeker. *There are three forms of radar seeker – semi-active, active and passive – and all home in on either reflected or emitted radar signals, deriving information about the target's position and velocity. This allows the missile to plot a course to intercept the target.*

carried by the launch aircraft, allowing reduced missile mass and power consumption, thus longer range and greater flight times. The transmitter can be of higher power and have a larger antenna because it is not limited by the size or available power of the missile. Disadvantages of this system are that changing transmission characteristics is more difficult, although not impossible. More importantly, the launch aircraft needs to keep the target illuminated, keeping it in the danger area whilst emitting a powerful transmission which may be easily detected by hostile forces. Semi-active systems are becoming less common as technology provides improved range with active systems.

By using the transmitted signal as a reference, both active and semi-active seekers can have the ability to deduce a variety of information about the target's position. Range can be calculated from the time taken for the reflected signal to return; range rate from Doppler shift; bearing and bearing rate from the 'look' angle and 'look'-angle rate of the seeker antenna. By effective use of this information, the missile can plot the most efficient route to intercept the target, rather than just flying straight towards it throughout the engagement.

Passive radar seekers are little used in isolation, except for anti-radar missiles designed to destroy radio and radar transmitters. However, many active and semi-active systems have a passive mode, called *home-on-jam* (HOJ). If the target emits a jamming signal in an attempt to divert the missile, the missile enters a passive mode, and uses the jamming signal itself to locate the target.

Infra-red systems almost invariably operate in the passive mode, that is, they do not transmit but receive the infra-red radiation that all bodies emit at temperatures above absolute zero. Because of this, the missiles themselves are difficult to detect, reducing or eliminating the warning that the target receives that it is under attack. Although there are many countermeasures against infra-red missiles, particularly flares which are intended to dupe the missiles into attacking them rather than the target aircraft, missiles are increasingly sophisticated in detecting and avoiding such distractions. Imaging infra-red (I^2R) goes one step further, creating an image composed of a large number of picture elements (pixels) which can be compared with pictures stored in memory to identify the target. Once this has been achieved, it is possible to select targets according to priority, rather than the brightest or nearest, or to configure the attack pattern to take advantage of inherent weaknesses. I^2R is used in many short-range weapons, including anti-aircraft, anti-armour and anti-ship systems. One disadvantage, however, is that infra-red and I^2R are not suitable for target ranges exceeding about 15 or 20 km, since infra-red signals are absorbed by the atmosphere.

Multi-mode seekers are now being developed which combine a radar seeker and an infra-red seeker into a single system. This can then offer either or both systems during any missile engagement.

Optical guidance systems are extensively used for ground-launched systems, but are rarely used for air-launched missiles.

Also involved in detecting and identifying the target is the interface between the launch aircraft and the missile before, during and after launch, transmitting information back and forth via an umbilical cable (before launch) and sometimes via a radio link in flight. Communications are obviously very important and transputers have greatly improved data-transmission rates, both within the missile and through the umbilical link.

Flight to the target

Launcher

In order for the missile to fly to the target, the aircraft must have some means of releasing its missile safely and effectively. This may be from a launcher rail –

a short track which holds the missile securely until the thrust from its propulsion system has built up sufficiently for it to disperse from the launch aircraft safely. Another commonly-used system is ejection launch, where the missile is pushed away from the aircraft by gas-operated rams, at acceleration levels around 15–20 'g' (see Figure 14.1). Rocket-motor ignition takes place immediately *after* launch, permitting semi-buried installation in a fuselage recess without damage from the motor efflux. Other missiles, particularly air-to-surface missiles which will not be launched during severe aircraft manoeuvres, may be simply released from the launcher, with the missile propulsion system ignited before or after launch depending on its type.

Motive power

A variety of power plants are used, depending on the type of missile. For many air-launched weapons, the solid-fuelled rocket motor is most commonly used, being reliable, easy to handle and safe, with a long, maintenance-free storage life. The thrust profile can be determined by the shape of the propellant charge to provide constant or varying thrust with time, and it can provide a relatively high thrust for a given weight and volume. Rocket motors often deliver extremely high thrust for a short duration, with the majority of the achievable flight time of many missiles occurring as coast (unpowered) flight. One disadvantage of the solid rocket motor over most other types of motor is that thrust cannot be controlled (or stopped) after ignition – because the charge is a mono-propellant (carrying its own oxygen), it burns at a pre-determined rate until exhausted. However, this limitation is outweighed by its advantages in many missile applications.

Gas-turbine engines can be used to provide a long thrust duration with relatively low thrust, making them highly suitable for long-range (over 80 km), subsonic flight. Since they are air-breathing engines, they do not have to carry their own oxygen, and so have reduced fuel consumption compared with rocket motors. This fuel economy is useful for anti-ship applications and cruise missiles. To allow launch at low or zero forward speed, helicopter-launched gas-turbine-powered missiles need a rocket booster for launch, which is jettisoned after burn-out. The thrust of a gas turbine can be controlled during flight by varying fuel flow, and it can be shut down if required.

Ramjets and pulse jets have limited use for missiles because of their performance limitations at low speed. Ramjets can be used by utilising a rocket motor to accelerate the missile to a speed at which it can operate, sometimes by putting the rocket propellant into the ramjet duct, which then takes over following rocket motor burn-out.

Control system

Although the seeker knows where the target is, the missile needs a control system to enable it to fly the course required. Some form of autopilot is required, unless the missile is controlled from the launch aircraft. The autopilot will take target bearing and position information from the seeker, and signals from on-board instruments such as gyroscopes and accelerometers to provide information on its own motion. The autopilot is programmed with the performance characteristics of the steering system, so that it can calculate the required steering commands.

Steering

Steering the missile can be achieved in a variety of ways. The most common is to use moving aerodynamic surfaces in a similar way to aircraft controls; these may be in a variety of configurations, such as fixed wings with moving tail fins, fixed fins with moving wings, fixed wings with moving canard surfaces, or fins only. The missile may either steer in the direction required, regardless of any sense of 'up', or it may bank and turn in the same way as an aircraft. For very short range 'dogfight' engagements, aerodynamic controls cannot achieve the very high turn rates required (sometimes exceeding 50 *g*), because of aerodynamic stalling. In this case, the rocket motor nozzle may be turned, deflectors impinged into the motor efflux or side thrusters used, to give *thrust-vector control*, or TVC. There are disadvantages with TVC – deflectors can suffer from very rapid erosion, and of course TVC only operates during motor burn, so this type of steering is generally limited to short-range missiles.

Power supply

Finally, the missile cannot operate without the required power supplies. Invariably, electrical supplies are required, but hydraulic power for operating control-surface actuators, and/or gas supplies for pneumatic control-surface actuators, cooling of infra-red detector elements or for pumping liquid fuels, may also be required.

Electrical power is generated by a thermal battery, which is a 'one-shot' device capable of delivering high power for a fairly short duration, with a very long, maintenance-free shelf life. It is initiated at launch, and generates power using an exothermic (heat-producing) chemical reaction under pressure. The electrical output is passed through a stabilising circuit, and is then split to supply the voltages and currents required by the various units. Before launch, power is usually taken from the launch aircraft via the umbilical cable.

Hydraulic energy is stored in a hydraulic accumulator, using pressurised nitrogen to store the pressure required, and initiated by firing a small pyrotechnic charge which bursts a diaphragm and allows pressure to reach the oil. There is no return line – unwanted hydraulic oil is ejected from the missile. Again, this system offers long shelf life with no maintenance.

Pneumatic energy may be stored as gas under pressure or generated chemically when required. Cooling of infra-red detectors requires a pure coolant gas such as nitrogen, but other applications such as pneumatic actuators are designed to tolerate the impurities present in gas which has been generated by a chemical reaction such as burning. Generating gases chemically is more practical when large amounts of gas are required.

Destruction of the target

Safety and arming unit

It is obviously essential that there is no possibility that the warhead can explode before or immediately after the missile is fired, endangering personnel or the host aircraft. Once the missile has been fired and is safely clear of the aircraft, the safety and arming unit (SAU) will arm, allowing firing signals to trigger the warhead. It does this by two means for fail-safety (if something fails the missile must always be left in a safe condition), one of which must normally be entirely mechanical. The first means is an electronic firing switch (EFS), which is operated by a timer as part of the launch sequence. Until the EFS is operated, the electrical firing signal is prevented from reaching the explosive charges which detonate the warhead. The second means is usually a clockwork mechanism. It arms by allowing the missile forward acceleration to move part of the mechanism so that the warhead initiator charges align with holes in the casing (called shuttering). In this way, even if the initiator charges in the SAU were to fire before the SAU had armed, the shuttering would prevent the explosion from detonating the main warhead charge. The missile must maintain a given acceleration for a minimum time for the SAU to arm, or the SAU will automatically return to its safe position. SAUs are very sophisticated mechanical items, made to close tolerances and extensively tested.

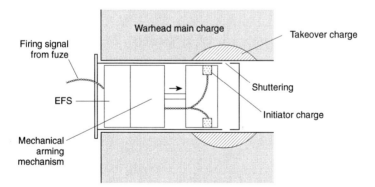

Figure 14.10 Principle of operation of safety-and-arming unit. *The electronic firing switch (EFS) prevents electrical firing signals reaching the explosive initiator charges until a pre-set time from launch has elapsed. The mechanical arming mechanism requires a sustained acceleration for a minimum time to move it to its armed position. In this position, the initiator charges are aligned with holes in the casing (shuttering), allowing the initiators to trigger the main charge, via the take-over charge, when the firing signal is received.*

Fuze

The purpose of the fuze (often spelt with a 'z', as here) is to detect when the missile hits the target, or when it makes its closest approach. Most anti-aircraft missiles do not need to hit the target, just to approach within a defined distance. All anti-aircraft warheads have a defined 'kill radius' – that is, the radius at which there is a given probability of causing sufficient damage to disable or destroy the target. This is determined by ground tests under controlled conditions. In some ways, the fuze operates in a similar way to the seeker, but with a different purpose. When the missile is in the optimum position relative to the target, a firing signal will be sent to the SAU, which will initiate the warhead.

Fuzes may be divided into two types – proximity fuzes and contact fuzes. Proximity fuzes, which detect a target from a distance, are usually radar or passive infra-red. They are designed to be insensitive to false signals such as electronic countermeasures, and, in the case of infra-red fuzes, flares and the sun. Because it is difficult to hit an aircraft because of its manoeuvrability and speed, proximity fuzes are widely used in anti-aircraft missiles. The warhead can inflict great damage to an aircraft from a distance of several metres, because of the vulnerability of aircraft compared with many other targets.

Contact fuzes operate only if the missile hits the target. Anti-armour missiles have contact fuzes, or laser fuzes to detonate the warhead at a pre-determined stand-off distance from the target. Anti-ship weapons also have contact fuzes, often with a delay to allow the missile to penetrate deep inside the target to maximise damage. Most anti-aircraft missiles have a contact fuze in addition to the proximity fuze, since a proportion of missile firings actually hit the target.

Warhead

The chemical compounds used to make up the explosive charge in warheads are changing dramatically at the present time. Improved performance is, of course, a prime aim, but many countries now have a requirement that explosives and propellants should be highly resistant to initiation by munitions. If missiles are struck by bullets or cannon fire, they must not explode or burn fiercely.

Warheads fitted to anti-aircraft and anti-armour missiles have very different characteristics, reflecting the differing requirements. With the exception of special warheads such as nuclear and chemical, there are three main types – blast, fragmentation and shaped-charge. There is also a common hybrid, the blast-fragmentation warhead.

Blast warheads use the very high pressures generated by the explosion to produce a blast wave which causes disruption to most structures. They can also be highly effective against personnel at short range. They are generally less effective than fragmentation warheads for many applications.

Fragmentation and blast-fragmentation warheads contain the pressure energy and then use it to break up the warhead casing, propelling fragments of the casing, and often additional pieces of material (together called *shrapnel*), at high speed. Close to the warhead, where there are many fragments close together, considerable damage may be inflicted on a wide range of targets. The shape of the charge is very carefully developed to give the required fragmentation pattern, since this is critical to kill radius. Blast-fragmentation warheads (Figure 14.11) typically consist of thousands of cubes, often of tungsten because of its high density, typically 1 cm across, packed tightly around the charge. On detonation, these cubes form a cloud of high-speed fragments, in a carefully designed pattern, accompanied by the blast from the charge detonation. The kill radius is defined by a specified minimum fragment density. This type is the most common in current use for anti-aircraft missiles.

Continuous-rod warheads (Figure 14.11) are a variation on the blast-fragmentation type, and consist of steel bars laid lengthways around the charge, welded at alternate ends. When the warhead explodes, these bars are thrown out, forming an expanding continuous ring, which acts like a circular saw. This gives a very high kill probability within the radius at which the ring begins to break up. Provided the missile can approach the target closely enough this is very effective, but once the ring reaches its full diameter and

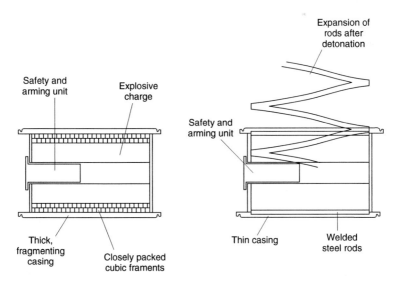

Figure 14.11 Blast-fragmentation and continuous-rod warheads. *The blast-fragmentation warhead generally consists of a large number of dense metal cubes packed around an explosive charge. The shape of the charge is carefully designed so that the fragments are ejected in a controlled pattern for maximum effect. The continuous-rod warhead, a variation on the blast-fragmentation warhead, is made up of a number of steel rods laid lengthways along the outside of the charge. The rods are welded at alternate ends, expanding in a single hoop which behaves rather like a circular saw.*

begins to break up, its effectiveness falls off quickly. One further advantage, however, is that target destruction is more easily seen on aircraft radar. Since a pilot must treat any hostile aircraft as a threat until there is clear evidence to the contrary, this can offer a tactical advantage.

Shaped-charge warheads (Figure 14.12) are specifically designed to attack armour, which is resistant to many other warhead types. Because the missile will actually hit the target, the entire effect of the warhead can be directed at one point. It is usually most convenient to direct this forward through the seeker, placing the warhead or warheads directly behind the seeker. The explosive is packed behind a copper cone, which focuses the effects to form a plasticised jet of copper travelling at very high speed and capable of penetrating a considerable thickness of armour plate. Time must be allowed for this penetrating jet to form, so detonation takes place *before* the missile reaches the target (known as *stand-off*). To cope with reactive armour (sheets of explosive placed on the

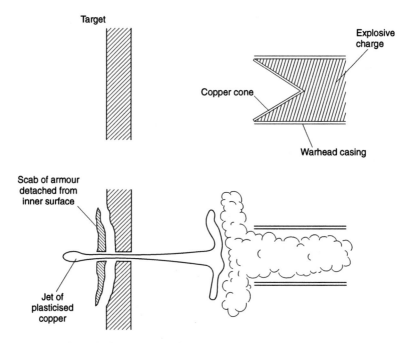

Figure 14.12 Shaped-charge warhead. *The shaped-charge warhead operates by focusing the action of the explosive into a copper cone, some of which then deforms into a narrow, fast-moving jet. This jet can penetrate great thicknesses of armour plating, provided it has time to form before impact. Firing the warhead before impact is known as stand-off, and may be achieved by a laser fuze or a mechanical trigger. Reactive armour disperses the effects of this warhead, but may be overcome by using two or more warheads together.*

outside of armour plate to disperse the explosive effect of warheads), multiple shaped charges can be used. The first charge initiates the reactive armour, leaving the standard armour exposed to a second charge in tandem. Where multiple layers of reactive armour are used, the warhead system will be effective provided there are more warhead charges than layers of reactive armour. Although there may be a seeker between the warhead and the target, penetration of the seeker causes little reduction in warhead effectiveness.

Location of parts

When designing the layout of an air-launched missile, a number of considerations must be taken into account. Besides the obvious requirement for the effective operation of certain units, tactical points may be relevant, such as whether the missile will be stored as a complete unit, or whether explosive components will be removed and stored separately. This may affect the location of some units. Invariably, the designer seeks the best compromise.

Seeker

Because the seeker needs an unhindered view of the target, it will normally be placed at the nose.

Control surfaces

Obviously, the position of these is dictated mainly by aerodynamic constraints, so the surfaces and their actuators will take high priority when allocating positions.

Fuze

Again, this is fairly important for anti-aircraft and anti-armour missiles, since the position of the fuze detector elements (aerials or windows) affects its response to target proximity.

Motor

An obvious place for a rocket motor would appear to be at the aft end of the missile, eliminating the need to build units around a blast pipe. However, rocket-motor charge accounts for 15–30 per cent of missile mass, and the centre of gravity of the missile will change during flight as the fuel is consumed, so the motor is sometimes placed more centrally to reduce its effect on the balance of the missile.

Warhead and SAU

Surprisingly, the warhead and SAU are among the least critical regarding position, except for anti-armour missiles, where the warhead would be close to the nose. However, the warhead is quite a heavy component.

Electronics, power supplies and umbilical

It is useful if the power supplies can be placed adjacent to the units using the power, to reduce cable lengths and improve maintainability. Hydraulic accumulators are usually integrated with control surfaces, to form one missile section; power supplies are sometimes integrated into electronic units, and cooling gas supplies are invariably built into seekers, and often have an additional gas connection from the launch platform via the umbilical.

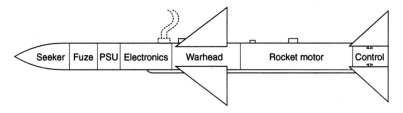

Figure 14.13 Typical internal layout of air-to-air missile. *This hypothetical example shows just one of a wide range of possibilities for the location of units within a missile.*

CHAPTER 15

THE COCKPIT

Objectives: to describe the features, instrumentation, environment and escape system of an aircraft cockpit.

INTRODUCTION

The cockpit is the point from which the pilot controls all of the operation of the aircraft. For a combat aircraft, the crew is confined to this area, so every single control necessary must be provided. In any aircraft, these controls must be easily reached, easily operated and logically arranged, so that the pilot instinctively knows where to find them, and the possibility of inadvertent or incorrect operation is minimised. All the information the crew need to carry out their duties must be presented in a clear and logical way, and no vital indication must be missed. As far as the pilot is concerned, the cockpit is the most important part of the aircraft. Since it is the pilot's workplace, it must also be as comfortable and safe as possible.

To meet all these requirements, the cockpit is very complicated, with every suitable surface used for instruments and controls. Although the cockpit of each aircraft type is different to every other, the general layout follows certain patterns, which helps crew to adjust quickly when they fly different aircraft types.

INSTRUMENTS

The greatest problem facing cockpit designers is to decide what information to show and in what format. At any time, there is a priority associated with each type of information, and the highest-priority information must be easily located and conveyed. If too much low-priority information is shown, the most important could be lost in a mass of data. However, if too little is shown, the crew may not have the information they need to fly the mission successfully.

Figure 15.1 Combat aircraft cockpit. *The layout of a combat aircraft cockpit is critical to the effectiveness of the aircraft. The layout must assist the crew to carry out their role effectively in the very demanding physical conditions a modern fighter creates. Every control and indicator must be logically and conveniently placed.*

The instruments which are in constant use are grouped together directly in front of the pilot, on the main instrument panel, with other instruments and controls arranged in groups to either side and on side consoles. Instruments which need to be seen by two or more crew members are duplicated, or may be placed centrally if the crew are seated side-by-side.

Whatever the aircraft, the pilot needs to know its attitude, and its position and speed in three dimensions. The instruments which provide this are placed centrally on the instrument panel. Generally, a standard configuration is used, as shown in Figure 15.3, although there are now many variations. The instruments are placed in the same positions on the instrument panel, although in some cockpit layouts the functions of two or more instruments can be combined. The so-called glass cockpit goes even further, with perhaps two display units performing the functions of all of these instruments. This is described later in this chapter.

The instruments needed to give basic information are:

- attitude indicator
- horizontal situation indicator
- air-speed indicator (ASI)
- Machmeter (high-speed aircraft only)
- altimeter
- vertical speed indicator

Figure 15.2 Commercial aircraft glass cockpit. *The instrument panel of a modern aircraft is very carefully designed to give the crew all the information they need to operate the aircraft effectively, but to present the information in the clearest way possible. The most important information must be located directly in front of the eyes, with colour coding and different forms of display used to allow the information to be taken in with only a brief look at the instruments. Any changes or deviations from the normal values must be easily recognised.* Photograph courtesy Jetstream Aircraft Ltd.

Figure 15.3 Basic instrument panel. *As a minimum, the instrument panel requires an airspeed indicator, altimeter and compass, but most designers would also include artificial horizon, turn-and-slip indicator and vertical speed indicator in this minimum specification. Engine instruments are also required. This photograph shows the second pilot's instrument panel from the Grand Caravan, and is a back-up panel only.*

Attitude indicator This indicates to the pilot the position and orientation of the horizon relative to the aircraft, and is also known as the artificial horizon (Figure 15.4). When visibility is obscured by cloud or at night, it is not possible to see the real horizon. The attitude indicator shows the angle of bank of the aircraft, and also the nose-up or nose-down attitude. It contains a gyroscope, which is rotating at high speed, and because of this will keep a constant orientation in space. By measuring this orientation relative to the aircraft, the instrument can indicate how much the aircraft's attitude has changed. A more modern version of attitude indicator is called an *attitude director*. It does the same job as the artificial horizon, but can also receive inputs from other systems, for example *ILS* (instrument-landing system), and give directions to the pilot to allow a particular flight path to be flown.

Horizontal situation indicator This does a similar job to a compass (in fact, it has a compass built into it), but also displays additional information (Figure 15.5). It displays VOR beacon information, as described in Chapter 12, on both the main compass rose scale and a digital read-out, and the distance of

Figure 15.4 Artificial horizon. *The artificial horizon shows the relationship between the aircraft attitude and the horizon. The fixed white symbol in the centre represents the aircraft, and the light and dark areas (actually blue and brown) the sky and ground respectively. The horizon moves vertically and also rotates as the aircraft manoeuvres, giving an artificial view of the real horizon in pitch and roll when visibility is poor.* Photograph courtesy Smiths Industries Aerospace.

Figure 15.5 Horizontal situation indicator. *This is an enhanced version of the simple magnetic compass. As well as the standard compass function, the instrument displays navigation information, such as VOR and DME, and also has an ILS (instrument-landing system) display.* Photograph courtesy Smiths Industries Aerospace.

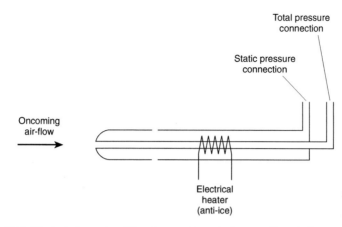

Figure 15.6 Pitot-static probe. *The pitot-static probe is open directly forwards, transmitting total pressure to the required instruments, and at right angles to the air flow, receiving static air, i.e. without the component due to the relative motion of the aircraft and the air. The difference between these two pressures is proportional to the square of the air speed at low speed. The heater prevents icing of the probe.*

the aircraft from a VOR/DME beacon. By combining these functions, the aircraft could have a VOR-DME-ILS receiver, which saves weight.

Air-speed indicator (ASI) Taking its input from the *pitot-static probe* (Figure 15.6), usually extending from the nose of the aircraft or wing leading edge, this instrument indicates aircraft speed relative to the air in which it is flying. It uses pressures from two parts of the probe. The static part of the probe is positioned so that the pressure it receives is the same as if the air were at rest relative to the aircraft. The pitot part is open directly forwards, and receives *ram air*. Ram-air or *total* pressure is the sum of static pressure and kinetic pressure. The kinetic pressure component is produced by bringing the air to rest, converting the kinetic energy of the air to pressure energy. The kinetic pressure (total pressure minus the static pressure) can be used to provide an indication of the speed, which is displayed by the ASI. However, it is important to realise that the ASI shows indicated air-speed (*IAS*), the effects of the compressibility of air introducing increasingly significant errors above Mach 0.3. Despite this, IAS is still very useful to the pilot, and a calculation to give true air-speed (*TAS*) can be done quite easily when needed, using either a hand computer or the aircraft's on-board computer. Air-speed is usually quoted in knots (nautical miles per hour).

Machmeter At higher speeds, it is more useful to the pilot to have speed information relative to the speed of sound, as a *Mach number*. The Machmeter is often combined with the ASI (Figure 15.7).

Altimeter Because ambient air pressure falls with altitude, the static pressure provided by the pitot-static probe can be measured and displayed on an

Figure 15.7 Combined air speed indicator and Machmeter. *Under various flight conditions, the pilot may be interested in either indicated air speed or Mach number. This instrument displays both simultaneously – IAS using the needle and dial, and Mach number via the digital display. Using different display modes eliminates confusion between the two.* Photograph courtesy Smiths Industries Aerospace.

Figure 15.8 Altimeter. *This altimeter uses a combined digital and analogue readout to indicate aircraft altitude. The digital display provides an unambiguous readout of altitude, while the analogue pointer gives a clearer indication of the rate of change of altitude. Since the altimeter is a barometric instrument, the reference pressure is set by the adjustment knob on the instrument, and is shown in the small windows (in this case 1013 millibar or 29.92 inches of mercury).* Photograph courtesy Smiths Industries Aerospace.

instrument calibrated to indicate altitude (Figure 15.8). Variations of ambient pressure as a result of meteorological conditions may be accounted for by adjusting the instrument, using the prevailing ambient pressure information provided by air traffic control (in millibar). The scale of the altimeter is always calibrated in feet, the standard used (almost) world-wide for defining aircraft altitude. In normal cruise, the pilot would use 1013 mb (standard sea level pressure) as the reference, giving instrument readings in height above sea level. When approaching an airfield, the pilot would set the local pressure at the airfield ground level, so that the altimeter would read height above ground level.

The accuracy of a barometric altimeter is not good enough to allow pilots to use this alone to show height if they need to carry out landings in poor visibility. In this case, the instrument landing system or a more accurate altimeter could be used. The *radar altimeter* (Figure 15.9) works by transmitting a radio pulse from the aircraft and measuring the time before its reflection from the ground is received. Although the time to be measured is extremely short, the radar altimeter is highly accurate. It will only give heights above local terrain because it works by reflecting radio waves off the ground.

Figure 15.9 Radar altimeter. *The radar altimeter displays height above ground level, using a reflected radar signal. It gives a very accurate indication, and is more useful than a barometric instrument at low level. Note that the display gives much greater precision at low altitudes, but only extends up to 5000 feet.* Photograph courtesy Smiths Industries Aerospace.

Figure 15.10 Vertical speed indicator. *The VSI provides an accurate indication of the rate of climb or descent. The position indicated in the photograph is level flight – movement of the pointer above the zero shows climb, in thousands of feet per minute, and below shows descent.* Photograph courtesy Smiths Industries Aerospace.

Vertical speed indicator: this instrument (Figure 15.10) shows whether the aircraft is climbing or descending, and how quickly. It has a zero position, showing level flight, at the 'nine o'clock' location, and is calibrated in feet per minute; if the needle is below zero it shows descent, and above the zero it shows climb.

THE GLASS COCKPIT

In both combat and transport aircraft, new aircraft designs are moving towards the 'glass cockpit' – the use of television-type Cockpit Display Units (*CDU*s) which can be switched to display a wide range of different information, in full colour. This is often known as *EFIS* – electronic flight-information system. For combat aircraft in particular, this means that a large number of individual instruments can be replaced by one or two screens. The information which is not needed during that part of the flight is not displayed, but the different display modes will show information which is relevant to the type or phase of mission that is being flown. For instance, a display could show a moving map, so that the aircraft location is kept fixed in the centre of the display and the map moves underneath it. When a weapon is selected, the display can then show all the information required to use that system.

Modern transport aircraft now use multi-mode LCD (liquid crystal display) or CRT (cathode-ray tube) displays as the main set of instruments, with 'real' instruments used for back-up only. The screens can also show other information, such as engine, navigation and fault data. They can also show maintenance data and previous fault information, and even the maintenance handbooks, for use by the ground maintenance engineers. Each display can be switched to provide the required information, and the source of data can also be switched. So in the event of a fault in the display itself, or in the systems that provide the information and processing to drive the displays, little inconvenience or hazard is presented, and multiple back-up facilities are provided. The glass cockpit system is extremely flexible, and can be set up to give the information in a way which is most easily read and understood by the crew.

HEAD-UP DISPLAY

During flight, especially in combat, a pilot needs to spend as little time as possible looking down at the instrument panel, and concentrate on what is happening outside. But the pilot still needs the information that the instruments provide. All combat aircraft have a way around this problem – the *head-up display* (HUD). The HUD (Figure 15.11) is made up of a projection system and a glass reflector, located above the instrument panel and directly in

Figure 15.11 Fast jet head-up display. *The head-up display, in this case from an F-16, provides instrument and navigation displays to the pilot without the need to look down into the cockpit. The image is focused at infinity, eliminating the need even to refocus the eyes to see the information.* Photograph courtesy GEC-Marconi Avionics.

the pilot's line of sight. The pilot can see through the glass, but it also reflects an image generated by the HUD projector showing the most important information from the instruments. The image is focused at infinity so that the information appears as part of the pilot's view at all times. As well as flight data, the HUD will show information such as gun sights, target marking and weapons data.

CONTROLS

Aside from the controls required to steer the aircraft in pitch, roll and yaw, many of the other systems need instructions from the pilot. Examples of these are controls for undercarriage retraction and extension; radio, navigation and avionics systems; weapon aiming and selection; fuel systems; oxygen and cabin conditioning systems; laser rangefinders. Each of these systems must have switches, buttons and indicators to allow their operation. The cockpit is laid out so that those controls which are least important, or are used least often, are tucked away, leaving the most accessible areas for the controls which the

Figure 15.12 Combat aircraft control column. *Modern fighter control layouts are based on the HOTAS concept (hands on throttle and stick), placing as many of the controls for important systems as possible on the throttles and control column. This allows pilots to operate main systems without taking their hands away from the primary controls.* Photograph courtesy GEC-Marconi Avionics.

pilot needs quick and easy access to. Some controls are combined with others, for instance the control column may carry switches and buttons to operate the radio, guns and weapon release. More switches may be added to the throttle levers, since the pilot's hands should be kept on these two controls as much as possible. The provision of all relevant controls within easy reach is based on the *HOTAS* concept (hands on throttle and stick), which is recognised as approaching the ideal situation to maximise pilot performance during combat.

ESCAPE SYSTEMS

Because of the speed that jet aircraft fly at, it is not possible for a pilot to jump out of an aircraft in an emergency – the force from the air stream is just too high. Aircrew rightly wish to be as safe as possible, whatever the situation. Almost all combat aircraft now have an *ejector seat* (Figure 15.13), often known as a 'bang' seat, which ejects itself and the crew member from the aircraft in an emergency. The seat is operated by pulling a lever or handle, setting into action a chain of events which is automatic from then on. The cockpit canopy is either jettisoned, or shattered by a small explosive cord embedded in it. Straps attached to the seat and the crew member's clothing shorten to prevent limbs striking parts of the cockpit. Then cartridges and rockets propel the seat up a rail and clear of the aircraft. A small parachute is deployed, which stabilises the seat against tumbling. At a safe height, the crew member's personal parachute is deployed, and the seat falls away.

The automatic operation of the ejector seat means that an injured pilot can be brought safely to earth even in a very weak or semi-conscious state. If a crew member of a two-seat aircraft is unconscious, and ejection is needed, a *command-eject* system means that one crew member can operate both ejector seats. The seats fitted to all modern aircraft are of the *zero-zero* type, meaning they can be used at zero speed and zero altitude – for instance if fire breaks out when the aircraft is at rest on the runway. Ejector seats have saved several thousand lives since they were first built into aircraft, and no doubt will save thousands more.

AIRCREW ENVIRONMENT

Because the aircraft flies at high altitude and, for combat aircraft, can create high accelerations, the natural environment is very hostile to crew and passengers. The low air pressure and low temperatures associated with flight at altitudes of up to 20 000 metres mean that some artificial means of simulating lower altitudes inside the cabin are required. Because of the low air pressure, insufficient oxygen can be supplied to the occupants; this can be overcome by pressurising the appropriate areas of the aircraft, by supplying supplementary oxygen via a mask, or both. The choice of method depends on the type of

Figure 15.13 Ejector seat. *Once initiated, an automatic sequence of events propels the seat and its occupant clear of the aircraft, and deploys the parachute at a safe altitude. Two-seat aircraft may have a command-eject system, which allows an unconscious crew member to be ejected by the other crew member. The zero-zero seat, used in all modern aircraft, can operate effectively with the aircraft at rest on the ground – the most adverse condition for operation.* Photograph courtesy Martin-Baker.

aircraft. For passenger aircraft, the fuselage will be pressurised, normally to a maximum equivalent altitude of about 2500 metres (8000 feet). Passengers will not normally be exerting themselves during the flight, so no discomfort is caused, and the structural weight penalty of pressurising to the equivalent of ground level is avoided. Combat aircraft will be pressurised to a higher equivalent altitude, typically 8000 metres (25000 feet). This is because a lower equivalent altitude (i.e. a higher level of pressurisation) carries a penalty in terms of higher structural weight to withstand the higher loads generated, and the effects of depressurisation, for example from battle damage, would be more serious. At this low degree of pressurisation, the crew need supplementary oxygen, which is supplied through a face mask. The same mask continues to provide oxygen if required during emergency escape. At extreme altitudes, without pressurisation, the oxygen must be supplied to the crew under pressure, in order to meet the body's needs. Oxygen is only needed in passenger aircraft if depressurisation occurs, usually due to some structural or pressurisation-system failure. In this event, oxygen will be supplied by face masks which drop automatically from the overhead lockers. The oxygen supply can be stored in the form of liquid oxygen, which is returned to gaseous form through a heat exchanger, or may be generated chemically.

The very high accelerations which military aircraft can attain puts extreme demands on the crew. Tight turns are an essential part of air combat, but are of little use if they cause the crew to lose consciousness. The main problem is black-out caused by reduced blood flow to the brain under high positive (i.e. pull-up) accelerations. The crew of many combat aircraft now use pressurised leggings, which inflate automatically when required, reducing the flow of blood to the legs and thus maintaining the supply to the brain. In negative *g* manoeuvres, the opposite problem occurs, where excess blood flow to the brain causes 'red-out'. Negative *g* manoeuvres are usually less violent, and less often used, but red-out occurs at lower accelerations than black-out, and preventative measures are more difficult.

Figure 15.14 F-16 canopy. *All-round visibility is important in combat aircraft, and the view provided by the F-16 canopy, plus a high seating position, is one of the best in current use. Rear-view mirrors, first used in the second world war, are also provided.*

VISIBILITY AND WINDOWS

The size of the cockpit windows, and the visibility offered to the crew, are inevitably a compromise depending on the type of aircraft, its role and the structural needs of the airframe. Civil aircraft do not need to offer all-round visibility. All that is required is a reasonable forward and sideways view from the aircraft, coupled with good visibility on the approach and for taxying. So windows will be kept reasonably small, minimising structural problems associated with pressurisation. Another advantage of small windows is that they are much less susceptible to damage from bird strikes.

All-round visibility is vital to fighter aircraft, because the majority of engagements are dog-fights, i.e. at close range. The pilot must be able to see potential targets, and also to have a clear view behind. The bubble canopy, as seen on the F-16, presents a very desirable solution, providing almost unlimited all-round vision (Figure 15.14). A high seating position for the pilot adds to the advantage. Since cabin pressure differentials are lower on a combat aircraft than on an airliner, the structural problems associated with large areas of transparency are less severe, but other problems are found, such as very high temperatures inside the cockpit on sunny days.

FUTURE DEVELOPMENTS

In the future, much of the progress that is made will be in electronic systems to improve the capability and performance of the aircraft. In particular, the glass-cockpit concept is likely to become more and more common, with most instruments being moved onto TV-type screens. Looking further ahead, the main instrument panel may become one single display, like a large computer screen, with every instrument needed being displayed automatically as a colour picture on the screen. Military crew will be able to operate every major control in the cockpit just by clicking a few buttons on the control column, never needing to take their hands off the control column and throttles.

Helmet-mounted sights are already being used, especially in helicopters. Projectors attached to the helmet direct an image of instrument data, and perhaps a weapon sight, onto the visor. This is an improvement on the HUD because it is visible in whatever direction the pilot is looking. The weapon can be aimed at the target just by turning the pilot's head so that the sight points at the target. This system is likely to be used more and more, because it is fast and easy to operate. With the advent of battlefield lasers, which can be used to blind aircrew temporarily or permanently, the cockpit of the future combat aircraft is likely to become closed to the outside world. The transparent cockpit canopy may disappear altogether. Helmets, similar to those being developed for 'virtual-reality' applications, will give the aircrew an artificial view of the outside, with superimposed cockpit displays and target information.

Figure 15.15 Helmet sights. *Helmet sights carry the concept of the head-up display one step further. The image is displayed inside the visor of the pilot's helmet, or in an eye-piece, rather than through a fixed lens in the cockpit. This allows the information to be viewed whatever direction the pilot is looking. Also shown here are night-vision goggles, which intensify low-light images to allow low-level operation at night.* Photograph courtesy Helmet Integrated Systems Ltd.

With airliners and general-aviation aircraft, the trends are towards better and simpler cockpit instruments. Modern instruments are becoming smaller, more flexible and more reliable. Multi-mode displays, based on computerised flight data systems, allow the information to be presented in a way which is optimised for readability, rather than being restricted by the capabilities of a mechanical instrument. Many instrument panels contain a number of identical display screens, each representing a different set of instruments. If one of the units fails, it is then quite straightforward to re-configure other instruments to display the most important information, and this would normally be carried out automatically by the system. The minor inconvenience of an instrument panel rearranged in this way is small compared with that of using small standby instruments located at the edges of the instrument panel.

INDEX